State and Local
Climate and Energy Program

LOCAL GOVERNMENT CLIMATE AND ENERGY STRATEGY SERIES

Combined Heat and Power

A Guide to Developing and Implementing
Greenhouse Gas Reduction Programs

Energy Efficiency

U.S. ENVIRONMENTAL PROTECTION AGENCY
2014

EPA's Local Government Climate and Energy Strategy Series

The Local Government Climate and Energy Strategy Series provides a comprehensive, straightforward overview of greenhouse gas (GHG) emissions reduction strategies for local governments. Topics include energy efficiency, transportation, community planning and design, solid waste and materials management, and renewable energy. City, county, territorial, tribal, and regional government staff, and elected officials can use these guides to plan, implement, and evaluate their climate change mitigation and energy projects.

Each guide provides an overview of project benefits, policy mechanisms, investments, key stakeholders, and other implementation considerations. Examples and case studies highlighting achievable results from programs implemented in communities across the United States are incorporated throughout the guides.

While each guide stands on its own, the entire series contains many interrelated strategies that can be combined to create comprehensive, cost-effective programs that generate multiple benefits. For example, efforts to improve energy efficiency can be combined with transportation and community planning programs to reduce GHG emissions, decrease energy and transportation costs, improve air quality and public health, and enhance quality of life.

LOCAL GOVERNMENT CLIMATE AND ENERGY STRATEGY SERIES

All documents are available at: *www.epa.gov/statelocalclimate/resources/strategy-guides.html*.

ENERGY EFFICIENCY

- Energy Efficiency in Local Government Operations
- Energy Efficiency in K-12 Schools
- Energy Efficiency in Affordable Housing
- Energy-Efficient Product Procurement
- Combined Heat and Power
- Energy Efficiency in Water and Wastewater Facilities

TRANSPORTATION

- Transportation Control Measures

URBAN PLANNING AND DESIGN

- Smart Growth

SOLID WASTE AND MATERIALS MANAGEMENT

- Resource Conservation and Recovery

RENEWABLE ENERGY

- Green Power Procurement
- On-Site Renewable Energy Generation
- Landfill Gas Energy

Please note: All Web addresses in this document were working as of the time of publication, but links may break over time as sites are reorganized and content is moved.

CONTENTS

Combined Heat and Power

EXECUTIVE SUMMARY

Developing and Implementing Energy Efficiency Programs

Saving energy through energy efficiency improvements can cost less than generating, transmitting, and distributing energy from power plants, and provides multiple economic and environmental benefits. Energy savings can reduce operating costs for local governments, freeing up resources for additional investments in energy efficiency and other priorities. Energy efficiency can also help reduce air pollution and GHG emissions, improve energy security and independence, and create jobs.

Local governments can promote energy efficiency in their jurisdictions by improving the efficiency of municipal facilities and operations and encouraging energy efficiency improvements in their residential, commercial, and industrial sectors. The energy efficiency guides in this series describe the process of developing and implementing strategies, using real-world examples, for improving energy efficiency in local government operations (see the guides on local government operations, energy-efficient product procurement, and water and wastewater facilities) as well as in the community (see the guide on affordable housing).

Energy Efficiency in Combined Heat and Power Production

This guide describes how local governments can lead by example and increase use of combined heat and power (CHP) in their facilities and throughout their communities. CHP, also known as cogeneration, refers to the simultaneous production of electricity and thermal energy from a single fuel source. This guide includes an overview of the benefits of CHP systems, costs, sources of funding, and case studies.

The guide is designed to be used by staff at local energy or environment agencies, local code enforcement officials and city planners, city councils, and mayors or county executives. It also provides information useful for

RELATED GUIDES IN THIS SERIES

- **Energy Efficiency:** *Energy Efficiency in Water and Wastewater Facilities*

  CHP systems are very compatible with wastewater treatment facilities that use anaerobic digesters. Anaerobic digesters produce a continuous flow of biogas that can be used as a fuel source. In addition, anaerobic digesters have a heat load small enough to be met by most CHP systems.

- **Renewable Energy:** *Landfill Gas Energy*

  Landfill gas, which consists of approximately 50 percent methane and 50 percent carbon dioxide, can be captured at municipal solid waste landfills and used as a fuel source for CHP systems.

- **Energy Efficiency:** *Energy Efficiency in Affordable Housing*

  Many local governments partner with private and nonprofit organizations to develop multi-family affordable housing. Through these affiliations, local governments can encourage developers to use CHP systems in multi-family housing units to increase energy efficiency and reduce costs.

- **Energy Efficiency:** *Energy Efficiency in K-12 Schools*

  A number of schools around the country are using CHP systems to reduce energy costs and improve energy supply reliability.

- **Energy Efficiency:** *Energy Efficiency in Local Government Operations*

  The use of CHP in government buildings and operations can help increase energy efficiency and reduce GHG emissions and criteria air pollutants by decreasing consumption of fossil fuel-based energy.

local government partners, such as local businesses, utilities, energy service companies, and non-profit organizations. Readers of the guide should come away with an understanding of options to improve energy efficiency using CHP, a clear idea of the steps and considerations involved in implementing CHP systems, and an awareness of expected investment and funding opportunities.

The guide includes descriptions of the benefits of CHP (section 2); opportunities to implement CHP systems (section 3); key stakeholders (section 4); strategies for promoting CHP projects (section 5); strategies for effective project implementation (section 6); costs associated with CHP systems and opportunities to manage these costs (section 7); federal, state, and other programs that may be able to help local governments with information or financial and technical assistance (section 8); and finally two case studies of CHP projects implemented at local government facilities (section 9). Additional examples of successful implementation are provided throughout the guide.

Relationships to Other Guides in the Series

Local governments can use other guides in this series to develop robust climate and energy programs that incorporate complementary strategies. For example, local governments could combine use of CHP with alternative fuel sources such as biogas generated at **wastewater facilities** or **landfill gas** captured at solid waste landfills to help achieve additional economic, environmental, and social benefits associated with reduced use of fossil fuels. In addition, because CHP systems require less fuel to produce the same output as conventional separate heat and power systems, use of CHP in government and community facilities helps increase energy efficiency and energy supply reliability while reducing costs.

See the box on page v for more information about these complementary strategies. Additional connections to related strategies are highlighted in the guide.

1. OVERVIEW

Combined heat and power (CHP), also known as cogeneration, refers to the simultaneous production of electricity and thermal energy from a single fuel source. Simultaneous production is more efficient than producing electricity and thermal energy through two separate power systems and requires less fuel. This reduction in fuel use can produce a number of benefits, including energy cost savings, reduced GHG emissions, and reductions in other air emissions.

CHP SYSTEM CONFIGURATIONS

CHP systems consist of a number of individual components—prime mover (heat engine), generator, heat recovery, and electrical interconnection—configured into an integrated whole. Every CHP application involves the recovery of otherwise-wasted thermal energy to produce useful thermal energy or electricity.

CHP systems can be configured either as a topping or a bottoming cycle, as explained below.

In a typical topping cycle system, fuel is burned in a prime mover such as a gas turbine or reciprocating engine to generate electricity. Energy normally lost in the prime mover's hot exhaust and cooling systems is instead recovered to provide heat for industrial processes (such as petroleum refining or food processing); hot water (e.g., for laundry or dishwashing); or for space heating, cooling, and dehumidification.

In a bottoming cycle system, also referred to as "waste heat to power," fuel is burned to provide thermal input to a furnace or other industrial process, and heat rejected from the process is then used to produce electricity.

The graphic below demonstrates the configuration of a typical topping cycle gas turbine CHP.

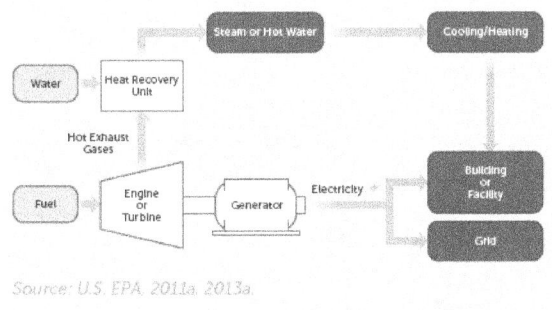

Source: U.S. EPA, 2011a, 2013a.

CHP systems are considered a type of distributed generation (also referred to as distributed energy and distributed energy resources), because they involve non-centralized, often small-scale projects. Distributed generation offers significant benefits, such as reduced risk of energy supply disruption, lower transmission and distribution losses, and reduced peak electricity demand on the grid. CHP systems produce the same benefits as those provided by typical distributed generation projects, but have the added benefit of increased energy efficiency (U.S. EPA, 2008b).

This guide provides information on how local governments have planned and implemented CHP systems at local government facilities and in their communities. It includes an overview of CHP system benefits, costs, sources of funding, and case studies. Additional examples and information resources are provided at the end of this guide in Section 10, *Additional Examples and Information Resources*.

2. BENEFITS OF COMBINED HEAT AND POWER

A CHP system appropriately sized to meet a facility's thermal energy needs achieves higher system efficiencies than conventional separate heat and power (SHP) systems that obtain their power and heat from different sources, such as central coal-fired power plants and onsite natural gas heating systems. Fossil-fueled power plants, for example, generally achieve a total system efficiency of approximately 33 percent, meaning 67 percent of the energy used to generate electricity is lost, often vented as heat. If a facility were to obtain its electricity through the grid from a fossil-fueled power plant and its thermal energy from an on-site boiler, the overall efficiency would be approximately 51 percent.[1] By using thermal energy that would otherwise be wasted in the power generation process, CHP systems can achieve total system efficiencies of approximately 60 to 80 percent (U.S. EPA, 2013b). The higher efficiency of CHP systems compared with SHP systems can help local governments and their communities to:

- **Reduce GHG emissions and other environmental impacts.** Because CHP systems require less fuel to produce the same energy output as SHP systems, CHP systems can reduce emissions of GHGs and air pollutants, such as nitrogen oxides (NO_x) and sulfur dioxide (SO_2). For example, a 5 MW CHP system with

---

[1] *This calculation assumes that the electricity and thermal energy used by the facility require 91 units of fuel at a central power plant (33% efficiency) and 56 units in an onsite boiler (80% efficiency). Transmission and distribution losses would further reduce overall efficiency. See http://epa.gov/chp/basic/efficiency.html.*

a natural gas turbine typically produces 20,870 metric tons of carbon dioxide (CO_2) emissions annually, while an SHP system designed to achieve the same output would produce 44,450 metric tons. The GHG emissions prevented by a CHP system of this size are equivalent to the annual GHG emissions of more than 5,400 passenger vehicles (U.S. EPA, 2013f).

The District General Services Plant in Sacramento, California, the state capital, was replaced in 2010 with a cogeneration plant that provides chilled water for cooling, steam for heating, and control air to 23 state-owned buildings in the Capitol Building district. The new plant uses 58 percent less energy and 90 percent less water than the plant it replaced, and its smaller footprint allowed the building team to reclaim land and turn it into a public garden, complete with a landscape feature that reuses water from the plant. The reduced water use allowed the plant to eliminate its former practice of discharging water into the Sacramento River, which was one of the major goals of the project. The new plant is Platinum LEED Certified. The reduced energy consumption of the plant avoids more than 3,900 metric tons of CO_2 annually, equivalent to the annual emissions from more than 800 cars. The plant is also equipped with combustion air preheating and low NO_x natural gas burner assemblies, which minimize combustion emissions of CO_2, carbon monoxide, NO_x, and ozone (Skanska, Undated).

- **Offset capital costs.** CHP systems can offset capital costs that would otherwise be needed to purchase and install certain facility components, such as boiler and chiller systems in new construction (U.S. EPA, 2013g). Installing CHP systems with backup capability can avoid the need for a local government to purchase a conventional backup electricity generator. A typical back-up diesel generator (with accompanying controls and switchgear) can cost as much as $550 per kW capacity (U.S. EPA, 2013d), compared with $100–$250 per kW to add backup capability to a CHP system (ORNL, 2013).

The wastewater treatment plant, in York, Pennsylvania, began operating a microturbine CHP system in 2011. A combined total of $4.2 million in state and federal funding allowed the project to move forward as scheduled, and reduced the impact of financing the improvements by the York City Sewer Authority's ratepayers. Sources of funding included $2 million in two separate awards through H_2O PA, $1.5 million from the Pennsylvania Alternative and Clean Energy program, and $500,000 from the Pennsylvania Department of Environmental Protection's GreenWorks program (BioCycle, 2012).

- **Support economic growth through job creation and market development.** Investing in CHP systems can help stimulate local, state, and regional economies. Demand for raw materials and for construction, installation, and maintenance services can create jobs and develop markets for CHP technologies (NECHP, 2006). Facilities that reduce their energy costs can spend those savings elsewhere, often contributing to the local economy (Lawrence Berkeley Laboratory, 2010).

Lansing, Michigan, has experienced an economic revival due to the local utility's installation of the REO Town gas-fired cogeneration plant. The installation of the plant coincided with the renovation of the adjacent Grand Trunk Western Railroad Depot, an historic landmark, which now serves as the utility's new headquarters. The new plant started operations in 2013 and serves the new headquarters. The construction process generated thousands of jobs and pumped an estimated $50 million in wages into the local economy. The revitalized downtown landscape has boosted the local economy, and is expected to attract new enterprises (MLive, 2013; Lansing, 2013).

Demonstrate leadership. Using CHP systems at local government facilities can be an effective and visible way of demonstrating environmental and fiscal responsibility to the public. Installing CHP systems at facilities frequently visited by the public can lead to greater community awareness of local government leadership and the benefits of clean energy.

The wastewater treatment plant in Sheboygan, Wisconsin, is recognized as a nationwide leader in energy efficiency in the water and wastewater treatment sector. The plant uses 20 percent less energy than its baseline level in 2003, and generates 70 to 90 percent of its own energy on site using a CHP microturbine system that runs on biogas produced in the wastewater treatment process. The biogas project, developed in a partnership between the local government and the local utility, allows the city to reap both the financial and societal benefits of producing renewable energy. The treatment plant receives one renewable energy certificate for every megawatt-hour of renewable energy the microturbines create (ACEEE, 2011). Renewable energy certificates represent the environmental attributes of electricity generated from renewable sources, and they

---

EPA CHP EMISSIONS CALCULATOR

The EPA CHP Emissions Calculator can be used to compare anticipated emissions from a CHP system with those from an SHP system. The calculator, which provides estimates for emissions of CO_2, SO_2, NO_x, N_2O, and CH_4, is designed for users with at least moderate familiarities with CHP technologies.

http://www.epa.gov/chp/basic/calculator.html

---

can be sold directly to customers or through brokers and marketers.[2]

Hedge against financial risks. The higher efficiencies of CHP systems relative to SHP systems translate into significant energy cost savings. Depending on total system efficiencies, a CHP system can consume up to one-third less energy than an SHP system (U.S. EPA, 2013b).[3]

EFFICIENCIES OF DIFFERENT CHP SYSTEMS

Not every CHP system operates at the same total system efficiency. A CHP system's efficiency depends on the technology used to generate the electricity and thermal energy, the system design, and how much of the thermal energy is used by the site. The most common prime movers include:

- Steam Turbine: 80% efficiency
- Diesel Engine: 70%-80% efficiency
- Natural Gas Engine: 70%-80% efficiency
- Gas Turbine: 70%-75% efficiency
- Microturbine: 65%-75% efficiency
- Fuel Cell: 65%-80% efficiency

Sources: U.S. EPA, 2013a; 2013b.

---

[2] *For more information on how local governments can demonstrate leadership by selling or purchasing renewable energy certificates and other forms of green power, see EPA's On-site Renewable Energy Generation and Green Power Procurement guides in the Local Government Climate and Energy Strategy Series.*

[3] *Based on a 5 MW natural gas-fired combustion turbine CHP system (U.S. EPA, 2013b).*

By installing a CHP system in 2008, the Back River Wastewater Treatment Plant in Baltimore, Maryland, reduced its electricity consumption by 19.4 million kWh annually. The system, which runs on methane gas produced in wastewater treatment, saved the City of Baltimore approximately $1.4 million per year in electricity costs, equivalent to 3.5 percent of the city's entire annual energy bill (Baltimore City DPW, 2012).

Since CHP systems require less fuel to produce the same output as SHP systems, they can help reduce the vulnerability of local governments to fluctuations in energy prices. Local governments can achieve additional protection from volatile energy prices by siting CHP projects in close proximity to biomass (e.g., wood and agricultural wastes) or biogas resources. Using renewable fuels can help the cost of operating CHP systems remain stable even as fossil fuel prices fluctuate. In addition, using biomass or biogas as a fuel source provides a use for material that often would be wasted otherwise (U.S. EPA, 2013f).

Lycoming County, Pennsylvania, partnered with a private renewable energy company to develop a 6.2 MW CHP system using landfill gas. This system supplies 80 percent of the Federal Bureau of Prisons' Allenwood Correctional Complex's electricity and 90 percent of the power and thermal needs of the Lycoming County landfill complex. The federal correctional facility gains long-term power price stability and clean energy to help it meet federal renewable energy requirements, and the county receives revenue for the landfill gas (PPL Renewable Energy, 2012).

- **Increase electricity reliability.** Disruptions in the energy supply can be a serious risk for local governments, many of which own facilities where a loss of electricity could be disastrous, such as waste water treatment facilities, hospitals, and schools. CHP systems can be designed to disconnect from the grid, enabling them to operate in "island mode" if grid-supplied electricity is lost during extreme weather or other circumstances, and provide increased reliability for these critical facilities. Using CHP to generate electricity on site also avoids the need to rely on non-CHP backup generators, and can even improve the overall reliability of the electricity grid by reducing peak load and reducing the risk of blackouts (U.S. EPA, 2013d).

During the New York City blackout in the summer of 2003 and Superstorm Sandy in 2012, half of city hospitals that used backup generators alone experienced failures (U.S. EPA, 2013d). In contrast, a study conducted after Superstorm Sandy showed that all of the CHP units in New York City designed to operate during a grid outage performed as expected when power was lost during the storm (NYSERDA, 2013). Hospitals, data centers, universities, district energy systems, and wastewater treatment plants with CHP systems were able to continue operating during and after the storm (ORNL, 2013).

In 2011, the Sonoma County (California) Administrative Campus installed a 1,400 kW fuel cell CHP system capable of providing 90 percent independence from the electric grid when needed. The CHP system is the centerpiece of the County's Comprehensive Energy Project, which the county expects will result in $40 to $50 million in energy cost savings over the next 30 years. The project will reduce and replace 13.4 million kWh of electricity use from the electric grid, reduce the county government's water consumption by 19 million gallons each year, and reduce GHG emissions by more than 5,400 metric tons annually, equivalent to the emissions from burning 600,000 gallons of gasoline (Sonoma County, 2011).

3. COMBINED HEAT AND POWER OPPORTUNITIES FOR LOCAL GOVERNMENTS

CHP systems can provide energy to be used in multiple applications, including power for facility operations and waste-heat recovery for facility heating, cooling, dehumidification, and other processes (U.S. EPA, 2013a). Several types of facilities and operations can be strong candidates for CHP systems, including:

- **Wastewater treatment facilities.** CHP systems are very compatible with wastewater treatment facilities using anaerobic digesters. Anaerobic digesters produce a continuous flow of biogas that can be used as a fuel source. In addition, anaerobic digesters require heat, which most CHP systems can provide. As of August 2013, CHP systems were operating at 185 wastewater treatment facilities in the United States, providing 612 MW of electric capacity. Of these sites, 143 were fueled by biogas generated at the facility, for a total of 352 MW (ICF, 2013). For more information on CHP in wastewater facilities, see EPA's *Energy Efficiency in Water and Wastewater Facilities* guide in the *Local Government Climate and Energy Strategy Series*.

GRESHAM, OREGON – CHP AT A WASTEWATER FACILITY

The City of Gresham installed a 395 kW CHP system at its 20 million-gallon-per-day wastewater treatment plant in 2005. The system, which is fueled using byproduct gas from the anaerobic digestion of sewage (consisting of 60% methane and 40% CO_2), provides more than 50% of the plant's electricity needs in addition to a substantial amount of the plant's heat. Annual energy cost savings are estimated to be approximately $200,000. For information on opportunities for CHP at wastewater treatment facilities, see *Opportunities for Combined Heat and Power at Wastewater Treatment Facilities: Market Analysis and Lessons from the Field* (U.S. EPA, 2011b) and EPA's *Energy Efficiency in Water and Wastewater Facilities* guide in the *Local Government Climate and Energy Strategy Series*.

Source: Energy Trust, 2006.

In 2009, the City of West Lafayette, Indiana, installed a CHP system in its wastewater treatment facility for the primary purpose of reducing GHG emissions. EPA recognized the project with a PISCES Award, which highlights projects that further the goal of clean and safe drinking water. EPA has also recognized the city as a Green Power Partner (Purdue University, 2010).

Landfill gas energy projects. Landfill gas (LFG), which consists of approximately 50 percent methane and 50 percent CO_2, can be captured at municipal solid waste landfills and used as a fuel source for CHP systems. LFG is generated in landfills continuously; it typically has a heating value of 500 Btu per standard cubic foot (scf), but values can range from 350 Btu per scf to 600 Btu per scf (U.S. EPA, 2013i). The economics of a landfill gas energy project (i.e., a project or facility capturing LFG for fuel use) improve the closer the landfill is to the end-user.[4]

In Los Angeles County, California, the Calabasas landfill operates three gas turbines that generate up to 10 MW of power for the local grid. The turbines, which were installed in 2010, have helped reduced NO_x and carbon monoxide to levels significantly below the mandatory limits in Southern California (Solar Turbines, 2011a).

In addition, LFG can be a green power source. Local governments can sell LFG to landfill gas energy project developers while retaining the environmental and technological attributes associated with the green power in the form of renewable energy certificates. For more information on landfill gas, see EPA's *Landfill Gas Energy* guide in the *Local Government Climate and Energy Strategy Series*.

K-12 schools. A number of schools around the country are using CHP systems to reduce their energy costs and improve reliability of their energy supply. Large schools, especially those with swimming pools and high hot water needs for cafeteria and locker room loads, are good candidates for CHP. Many schools also serve as emergency shelters during extreme weather events, and the CHP system can help the school remain operational during power outages.

Cambridge Rindge and Latin School, a public school in Cambridge, Massachusetts, installed a gas-fired cogeneration system in 2011 as part of a broader set of energy-saving improvements. The system powers the Media Arts building and provides heat and hot water to the school in winter. In summer, the school uses the cogeneration unit to heat the pools at the adjacent Cambridge War Memorial Recreation Center. The school anticipates the energy savings from the overall energy improvement project will reduce the its utility bills by more than $335,000 a year (Cambridge Public Schools, Undated).

For information on how K-12 schools can improve their energy efficiency, see EPA's *Energy Efficiency in K-12 Schools* guide in the *Local Government Climate and Energy Strategy Series*.

---

4 *The piping distance from a landfill gas energy project to its end-user is typically less than 10 miles, although piping LFG up to 20 miles can be economically feasible, depending on gas recovery at the landfill and energy load at the end-use equipment (U.S. DOE, Undated).*

Multi-family housing. Many local governments partner with private and non-profit organizations to develop multi-family affordable housing. Through these affiliations, local governments can encourage developers to use CHP systems in multi-family housing units. In some local governments, public housing authorities have taken the initiative of installing CHP systems in the facilities they manage.

In 2012, the Reading (Pennsylvania) Housing Authority installed a 400 kW CHP system at Glenside Homes, a 400 unit multi-family residential building. The housing authority was able to take advantage of energy efficiency and renewable energy federal grant opportunities to fund the project, which has an estimated annual energy cost savings of $75,000–$100,000 (UGI Performance Solutions, 2012).

USING CHP IN MULTI-FAMILY AFFORDABLE HOUSING

The U.S. Department of Housing and Urban Development (HUD) has developed guidance and a tool for using CHP in multi-family affordable housing. The guidance materials, which include a question-and-answer guide, a feasibility screening guide, and a library of resources, are available at *http://www.hud.gov/offices/cpd/library/energy/index.cfm* and *http://www.hud.gov/offices/cpd/energyenviron/energy/library/#chp*.

HUD has also developed a screening tool for CHP in multi-family housing. Users enter 12 months of energy use data, which the tool uses to calculate the potential savings and payback period from using a CHP system instead of obtaining heat and power separately. The tool is available at *http://www.hud.gov/offices/cpd/library/energy/software.cfm*.

District energy systems. In a district energy system, steam, hot water, or chilled water is generated at a central plant and then piped to many facilities throughout the district. These systems reduce energy consumption and capital costs by eliminating the need to install chillers and boilers in individual facilities. District energy systems can incorporate CHP systems into their configuration, leading to greater energy efficiency (IDEA, 2013; IDEA, 2007; IEA, 2013).

St. Paul, Minnesota, contracts with a private district energy provider to obtain hot water and cooling for many of its local government facilities. The district energy system relies on a 25 MW CHP system to generate energy. St. Paul has formed an agreement allowing the district energy provider to obtain approximately 300,000 tons of wood waste from the city's recycling center to be used to fuel the CHP system (St. Paul, 2013).

4. KEY PARTICIPANTS

A variety of participants can play important roles in mobilizing resources and ensuring effective implementation for CHP projects at local government facilities, including:

Mayor or county executives. The mayor or county executive can play a key role in increasing public awareness of the benefits of CHP. Including CHP goals in a mayor's or county executive's priorities can lead to increased funding for CHP potential studies and projects.

City and county councils. Clean energy activities, including efforts to increase use of CHP, are often initiated by city and county councils. In many local governments, city or county councils must authorize large capital expenditures, such as purchasing CHP systems. Securing support from city or county council members can be important for ensuring CHP initiatives receive the resources necessary to produce results.

The Orange County Cogeneration Plant is a 10.4 MW system serving the Santa Ana Civic Center Campus in Santa Ana, California. Installed in 2009 and powered by two gas turbines, the system provides electricity, heating, and cooling to buildings on the campus. Though economics were an important factor in the city's decision to install a CHP system, city officials have cited reliability and emissions reductions as key drivers in their decision to switch to CHP (Combined Cycle Journal, 2010).

Local code enforcement officials and planning departments. In some local jurisdictions, older zoning, building, and fire codes can present barriers to CHP project permitting. Local governments can work with their code enforcement officials and planning departments to update codes to accommodate CHP projects (Virginia DEQ, 2004). Some local governments, such as Boston, Massachusetts, and Epping, New Hampshire, have modified zoning ordinances to provide permitting incentives for CHP projects. Planning departments can also be responsible for developing local energy plans that can include CHP-specific goals and activities.

- **State energy and environmental departments.** State energy and environmental offices can provide local governments with information resources and technical assistance in planning and permitting CHP systems for local government facilities.

> In Lancaster County, Pennsylvania, the county partnered with the Pennsylvania Department of Environmental Protection to install a 3.2 MW CHP system that uses methane produced at two local landfills (U.S. EPA, 2013h).

- **State public utilities commissions (PUCs).** Local governments can work with state PUCs to obtain information on connecting CHP systems to the electricity grid and to learn about funding opportunities available for CHP projects. Some state PUCs administer programs that offer clean energy options for a targeted customer base or provide financial incentives for distributed generation projects, including CHP.

> State legislation enacted in 2009 directed the Maine Public Utilities Commission to develop a program offering green power (including CHP) as an option to residential and small commercial customers. The PUC issued rules in October 2010, and selected a company to manage the statewide green power program for Maine's transmission and distribution territories. The program, which began operating in April 2012, includes community-based renewable energy projects to the extent possible (Maine Public Utilities Commission, Undated).

- **Local businesses.** Local governments can contract with local businesses to provide electricity or thermal energy generated by local government CHP systems, or to purchase the energy generated by a privately

owned CHP system. Local industries can also provide feedstocks or fuels for CHP systems operated by local governments.

> The Gloversville-Johnstown Joint Wastewater Treatment Facility, in Johnstown, New York, works with several nearby cheese and yogurt producers to collect their wastewater, which the facility processes to generate biogas and then burns in a CHP plant to meet its electricity and heating needs. The arrangement provides the dairy companies with an economical disposal method for their wastes, while bringing revenue and a significant supply of biogas to the treatment facility. The CHP plant avoids 1,950 metric tons of CO_2 emissions per year, equivalent to the annual GHG emissions of more than 400 passenger vehicles (Cogeneration and On-Site Power Production, 2011).

- **Utilities.** Local governments can work to connect CHP systems to the grid either through utilities they own or by working with utilities with service territory in their area. Connecting to the grid allows local governments to help meet electricity loads when demand exceeds the capacity of the CHP system, and can create opportunities for local governments to sell electricity to the utility when capacity exceeds demand and where state interconnection and net metering rules permit (see Section 6, *Strategies for Effective Project Implementation* for more information on selling excess electricity). Information on state interconnection and net metering rules, which determine whether and how a utility allows customers to connect to the grid, can be accessed through the EPA CHP Partnership website (*http://epa.gov/chp/policies/database.html*). In addition, many utilities offer financial incentives for CHP projects through energy conservation programs.[5]

> A number of municipally owned electric utilities, such Gainesville Regional Utilities in Gainesville, Florida, are taking advantage of the efficiency benefits of CHP systems to provide customers with clean energy while reducing the utilities' operating costs (GRU, 2010).

---

5 *The Database of State Incentives for Renewable Energy provides information on utility incentives for CHP (http://dsireusa.org/index. cfm?&CurrentPageID = 2&EE = 1&RE = 0).*

Philadelphia Gas Works (PGW) promotes microturbines and other types of CHP projects as a way to sell more natural gas. As a landlocked utility serving a city whose population and industrial base have declined, PGW has a lot of underused infrastructure. With the price of natural gas low, partly because of an abundance of fuel produced from shale formations, PGW highlights the savings of natural gas over electricity or steam. To induce customers to invest, PGW offers a discounted industrial cogeneration rate for CHP projects that recover at least 15 percent of the waste heat (Philadelphia Inquirer, 2011).

- **Energy service companies (ESCOs).** ESCOs provide technical expertise on energy efficiency projects and often offer performance contracts, which typically include a guarantee that payments will not exceed the savings generated. Local governments can contract with ESCOs to purchase and install CHP systems and to obtain operations and maintenance services. For more information on performance contracting, see Section 7, *Costs and Funding Opportunities*.

In 2009, the Southeastern Regional School District in Easton, Massachusetts, installed a 250 kW reciprocating engine-powered CHP system after entering into an energy service performance contract with an ESCO. The performance contract, which includes a broad range of energy and water conservation projects in addition to the CHP system, is expected to save the district $276,000 annually (Ameresco, 2011).

- **Non-profit organizations.** Local governments can work with non-profit organizations to obtain technical or financial assistance for implementing CHP-related activities.

The International District Energy Association promotes energy efficiency and environmental quality through the advancement of district heating, district cooling, and cogeneration, and actively lobbies to secure favorable policies, legislation, and regulations for district energy. For example, the association has worked with brownfield business development parks that are evaluating CHP as an option.

5. FOUNDATIONS FOR PROJECT DEVELOPMENT

Local governments use several mechanisms to promote CHP systems in local government facilities and throughout the community, including:

- **Local government planning processes.** A number of local governments have included goals for CHP in local planning documents.

In its 2007 Climate Change Local Action Plan, the City of Philadelphia, Pennsylvania, recognizes the role CHP can play in reducing GHG emissions from city buildings. The plan includes an element for buildings intended to promote the installation of CHP at city complexes. Initially, the city will evaluate the financial feasibility of CHP at its prison system complex, and will then explore approaches to encourage CHP at other public and private facilities (Philadelphia, 2007).

- **Modification of zoning ordinances.** Local governments can encourage using CHP systems by modifying zoning ordinances.

Boston, Massachusetts, adopted a Green Building Ordinance in January 2007 that applies to new public and private buildings of 50,000 square feet or greater. To encourage CHP, the city included an additional provision awarding one credit point toward LEED certification for buildings drawing 10 percent of their total energy use from CHP systems (Boston, 2007a).

- **Incentives for private and public entities.** In addition to installing CHP systems at local government facilities, local governments can work with the private sector to encourage CHP systems in industrial manufacturing facilities, institutions, commercial buildings, and multi-family housing complexes. Local government activities to encourage CHP in the private sector include:

  - *Financial incentives.* A number of local governments offer financial incentives to local businesses and residents to install CHP systems at their facilities. These incentives are included in the Combined Heat and Power Partnership CHP Policies and Incentives Database (see text box on page 12).

    In 2003, Chicago, Illinois, organized a seminar to provide local hospital administrators with information on how CHP technologies can be applied cost-effectively at hospitals. The city followed this seminar with an offer to fund 50 percent (up to $5,000) of the cost of a CHP screening analysis to provide estimates of the costs, savings, paybacks, and internal rates of return for each participating hospital (Chicago, 2008).

  - *Outreach.* Some local governments are encouraging local businesses to install CHP systems at their facilities through educational outreach.

    Kauai County, Hawaii, has offered technical workshops through its Energy Extension Service to advise local businesses of opportunities to use CHP systems (U.S. EPA, 2008a).

6. STRATEGIES FOR EFFECTIVE PROJECT IMPLEMENTATION

Local governments can use a number of implementation approaches to enhance the benefits of CHP-related activities, including:

- **Assess local CHP potential.** An accurate initial assessment of local demand, potential barriers, and the availability of fuels can help ensure successful implementation of CHP projects. Local governments can obtain assistance from a number of resources to help assess CHP potential. For example, EPA has developed a CHP Project Development Handbook local governments can use as a guide throughout the five stages of developing a CHP project, from initial assessment to operations and maintenance (for more information on these stages, see the CHP Project Development text box on page 10). Some local governments have hired consultants to assess local potential for government, commercial, and residential CHP systems.

  San Francisco, California, commissioned a study to evaluate CHP potential across the city. The study identified potential demand for CHP in the city, assessed CHP equipment supply in the region, and identified local installation concerns, such as site selection and permitting issues (San Francisco, 2007).

- **Select an approach to project development.** Some local governments have chosen to self-develop CHP projects, hiring consultants to help plan and manage the design and construction process. This option can maximize financial returns to the local government, but involves more risk and requires significant personnel resources. Other local governments have purchased "turnkey" CHP systems that are planned, designed, and constructed by private developers that provide a single point of contact. This approach shifts some risk to the developer but may reduce the economic return to the facility owner. Local governments can also team with partners (e.g., an engineering firm) to develop the project and share the financial returns as well as the risks (U.S. EPA, 2013j). The EPA Combined Heat and Power Partnership developed a guidance document that provides information on determining whether to

CHP PROJECT DEVELOPMENT

Planning, installing, and operating CHP systems requires attention to many implementation issues. These issues can be addressed in five steps:

- **Qualification assessment.** At the initial planning stage of project, it is important to determine whether CHP is appropriate for the site in consideration. Many technical and economic factors must be considered, such as electricity and thermal energy demand and source fuel availability.

- **Level 1 feasibility analysis.** Goals at this stage include identifying project goals and potential barriers, and quantifying technical and economic opportunities.

- **Level 2 feasibility analysis.** Goals at this stage include optimizing CHP system design, accounting for capacity, thermal output, and operation needs. This stage should also involve final CHP system pricing and a determination of expected investment return.

- **Procurement.** This stage involves selecting a qualified contractor or developer, financing the project, and ensuring and recording compliance with siting and permitting requirements.

- **Operations and maintenance.** This stage involves maintaining a CHP system so that it continues to provide expected energy savings and emission reductions.

The EPA Combined Heat and Power Partnership has developed tools and resources, including a Spark Spread Estimator and a CHP Project Development Handbook, to assist with each of these stages in the CHP project development process. Local governments can find these resources on the CHP Partnership Website: *http://www.epa.gov/chp/project-development/index.html*.

Source: U.S. EPA, 2013h.

hire or partner with a developer, and how to select one that is qualified.[6]

Austin Energy, the local municipal energy utility in Austin, Texas, worked with a private turnkey CHP developer to create a CHP plant for the Dell Children's Medical Center of Central Texas in 2006. The medical center saved $8 million in capital outlay by outsourcing its power, heating, and chilled water needs to Austin Energy, which owns and operates the plant. The medical center will purchase its power and chilled water from the CHP plant at tariffed rates over a 30-year term (TAS, Undated).

- **Enter maintenance contracts.** CHP systems involve complex components that need to be maintained in order to continue working as designed. Local governments can enter maintenance contracts with equipment manufacturers and ESCOs for regular maintenance and operations services on CHP systems (U.S. HUD, 2005).

Millbrae, California, incorporated provisions into an energy performance contract with an ESCO to have future maintenance and potential renovations performed on a new CHP system (Chevron, 2006; Millbrae, 2005).

- **Involve local planning departments.** Local government planning departments typically need to verify that CHP projects are consistent with local land use and zoning regulations. In some localities, special use permits might be required for the construction of CHP systems.

The San Jose, California, Department of Planning, Building, and Code Enforcement developed a set of instructions for obtaining a special use permit to construct CHP systems (San Jose, 2008).

CONNECTING TO THE GRID

Local governments can connect to the grid to obtain electricity to supplement power produced by a CHP system, and to sell excess electricity to a local utility or provider. Interconnection rules vary by utility, but a number of states have adopted standardized rules making interconnection more streamlined. The EPA CHPP Procurement Guide includes information on the steps involved in establishing interconnection (see text box on page 12).

The Interstate Renewable Energy Council tracks state interconnection regulation activities. A table of these activities is available at *http://www.irecusa.org/fileadmin/user_upload/ConnectDocs/December_2007_IC_Table.doc*.

---

6 See *http://www.epa.gov/chp/documents/pguide.pdf*.

• **Sell excess energy.** In states where interconnection and net metering rules permit, local governments can sell some or all of the electricity they generate from CHP systems. To learn about state interconnection and net metering rules, visit dCHPP, EPA's database on CHP policies and incentives, at *http://www.epa.gov/chp/policies/database.html*.

> Winnebago County, Wisconsin, earns between $400,000 and $500,000 annually from selling electricity generated by a 1.06 MW CHP system at the county sheriff's office to the local electric utility. The office retains the 4,700 MBtu per hour of thermal energy for space heating and domestic hot water (Winnebago, 2007).

7. COSTS AND FUNDING OPPORTUNITIES

CHP systems involve a significant financial commitment. Fortunately, many funding opportunities are available for purchasing and installing CHP systems. This section provides an overview of the costs associated with CHP systems and opportunities to manage these costs.

Costs

The actual cost of a CHP project varies depending on a number of characteristics, including who develops the project (i.e., the local government or a private developer as part of a "turnkey" arrangement), system capacity, availability and type of fuel, prime mover, and overall system configuration. A typical CHP system can cost between $1,000 and $5,000 per kW of installed capacity (CEC, 2012). In addition to the cost of purchasing and installing the system, a CHP project will incur other associated costs in conducting preliminary feasibility studies, obtaining permits, and operation and maintenance costs. Preliminary feasibility studies, for example, can range from $10,000 to $100,000 (U.S. EPA, 2013l), and operations and maintenance costs can range from $0.005 per kWh to $0.015 per kWh (U.S. EPA, 2013k). EPA's CHP procurement guide, which discusses various financing methods for CHP projects, can be found at: *http://www.epa.gov/chp/documents/pguide_financing_options.pdf*.

In addition, EPA's CHP Catalog of Technologies (*http://www.epa.gov/chp/technologies.html*) provides information on cost and performance characteristics for the five key CHP prime movers.

Funding Opportunities

Many funding opportunities are available to local governments to help finance CHP installations, including:

• **Performance contracting.** Several local governments have used energy performance contracts to purchase, install, and maintain CHP systems. An energy performance contract is an arrangement with an ESCO that bundles together various elements of an energy-efficiency investment, such as installation, maintenance, and monitoring of energy-efficient equipment. These contracts, which often include a performance guarantee to ensure the investment's success, are typically financed with money saved through reduced utility costs but the systems may also be financed using tax-exempt lease-purchasing agreements.

MILLBRAE, CALIFORNIA – ENERGY PERFORMANCE CONTRACTING

Millbrae, California, used an energy performance contract to install a 250 kW methane-fueled microturbine CHP system at its wastewater treatment facility. The ESCO also integrated a grease receiving station into the facility to receive waste grease from local businesses, providing the CHP system with an additional source of fuel. The city saves $112,000 per year in energy costs by producing heat and power on-site, and the grease receiving fees generate $152,000 annually. These savings and income are used to pay for the system, which was installed at no upfront cost. This integrated system reduces the city's annual CO_2 emissions by approximately 544 metric tons.

Source: Chevron, 2006.

• **State government programs.** These programs offer financial incentives for distributed generation projects (including CHP) and projects using biomass/biogas (including CHP), and CHP qualifies as renewable energy in the renewable portfolio standards of 10 states. The most frequent state incentives include tax credits, rebates, and low-interest loans for CHP including biomass/biogas projects. Some states, such as California, Connecticut, and New Jersey, have included CHP as a critical component of their state energy strategies.

New Jersey provides grants for CHP systems based on system size and amount of electricity the system generates. The grant award cannot exceed $3 million per project, and the maximum grant percentage of a project's total cost is capped at 45 percent for fuel cells and 30 percent for all other CHP. The funds are made available through the New Jersey Economic Development Authority (U.S. EPA, 2013m).

In 2011, a biogas-fueled CHP system went online at the Beaver Dam Wastewater Treatment Plant in Beaver Dam, Wisconsin. The $20 million project was funded through a $10 million low-interest Clean Water State Revolving Fund loan from the government of Wisconsin and a $10 million grant from the American Recovery and Reinvestment Act. The plant uses whey waste from a local cheese factory to produce biogas and electricity, which is sold to a utility, as well as heat for the buildings at the plant (Renewable Energy From Waste, 2013).

- **Federal government.** Local governments can obtain financial assistance for CHP projects from the federal government. The EPA Combined Heat and Power Partnership maintains a database of federal and state financial assistance opportunities for CHP projects (see text box below). The database also provides information on non-financial incentives, including favorable regulatory treatment.

- **Utilities.** Some utilities offer financial assistance to local governments to help cover the costs of purchasing and installing CHP systems. A number of utilities also offer rebates for CHP systems.

The East Bay Municipal Utility District, a publicly owned utility serving two counties in California, received a $900,000 rebate from an investor-owned utility through the state's Self-Generation Program. The municipally owned utility used the rebate to offset the $2.5 million cost of a new 600 kW CHP system (EBMUD, 2006).

- **Non-profit organizations.** Local governments can often obtain financial assistance for CHP projects through non-profit organizations promoting clean energy. Most non-profit assistance for CHP takes place through project development support, in which the non-profit lends its expertise to evaluating CHP project feasibility. The non-profits typically receive their funding through federal or state agencies.

In 2013, the U.S. Department of Energy awarded a four-year, $2.2 million grant to the Southwest Energy Efficiency Project (SWEEP), a regional non-profit organization, to promote energy efficiency through the use of CHP. Based in Boulder, Colorado, SWEEP will promote best practices for CHP project financing, management and state policies, market analysis tools and resources, and technical site evaluations with businesses and communities in Colorado, Arizona, Oklahoma, New Mexico, Texas, Utah, and Wyoming (SWEEP, 2013).

8. FEDERAL, STATE, AND OTHER PROGRAM RESOURCES

Local governments can obtain information on developing CHP projects and about CHP systems in general through many federal, state, and other programs.

Federal Programs

- **U.S. EPA Combined Heat and Power Partnership.** The EPA Combined Heat and Power Partnership is a voluntary program that seeks to reduce the environmental impact of power generation by promoting the use of highly efficient CHP. The Partnership works with clean energy stakeholders from the private and public sectors to support the deployment of new CHP projects and to promote their energy, environmental, and economic benefits. CHP systems using at least 10 percent less fuel than comparable separate heat and power generation can qualify for the ENERGY STAR® CHP Award.

Website: *http://www.epa.gov/chp/*

- **U.S. EPA State and Local Climate and Energy Program.** This program helps state, local, and tribal governments achieve their climate change and clean energy goals by providing technical assistance, analytical tools, and outreach support. It includes two programs:

- *The Local Climate and Energy Program* helps local and tribal governments meet multiple sustainability goals with cost-effective climate change mitigation and clean energy strategies. EPA provides local and tribal governments with peer exchange training opportunities along with planning, policy, technical, and analytical information that support reduction of GHG emissions.

- *The State Climate and Energy Program* helps states develop policies and programs that can reduce GHG emissions, lower energy costs, improve air quality and public health, and help achieve economic development goals. EPA provides states with and advises them on proven, cost–effective best practices, peer exchange opportunities, and analytical tools.

Website: *http://www.epa.gov/statelocalclimate/*

- **U.S. DOE Industrial Distributed Energy Program.** The DOE Industrial Distributed Energy Program focuses on deployment of innovative CHP solutions through the support of technology development efforts, demonstrations, and technology performance validation projects conducted through public-private collaborations. DOE has also established several regional CHP application centers (now referred to as CHP Technical Assistance Partnerships) across the country. These centers can provide information on the benefits of CHP systems and project-specific support, including feasibility screenings and third-party reviews of vendor proposals.

Websites:

http://www1.eere.energy.gov/manufacturing/distributedenergy/ (Distributed Energy Program)

http://www1.eere.energy.gov/manufacturing/distributedenergy/chptaps.html (CHP Technical Assistance Partnerships)

- **Oak Ridge National Laboratory (ORNL) Cooling, Heat, and Power Technologies Program.** ORNL works with industry and the federal government to develop CHP technologies. The CHP Program can provide local governments with information resources on assessing site feasibility, evaluating CHP system benefits and performance, and capacity optimization.

Website: *http://www.ornl.gov/sci/ engineering_science_technology/ cooling_heating_power/*

- **U.S. Department of Housing and Urban Development (HUD).** HUD administers a number of programs intended to improve energy efficiency in the nation's public housing. Through these programs, HUD provides information on energy efficiency measures that can be implemented in multi-family developments, along with financial assistance for local governments and public housing authorities. The HUD Office of Energy and Environment, for example, provides information on CHP systems in multi-family housing. HUD, DOE, and EPA have recently issued a guide on CHP, reliability, and resiliency: *http://portal.hud.gov/hudportal/HUD?src=/program_offices/sustainable_housing_communities/chpguide*.

Website: *http://www.hud.gov/offices/cpd/library/energy/index.cfm*

State Programs

Some states have developed programs to promote CHP and other distributed generation technologies. Local governments can look to these programs for information resources on the benefits and applicability of CHP systems, as well as information on available financial assistance.

The California Energy Commission has developed a distributed energy resource guide that includes information on CHP systems, including cost ranges, efficiency performance estimates, relative strengths and weaknesses of CHP technologies, and lists of vendors (CEC, 2012).

New Jersey is seeking to improve its energy resilience through the New Jersey Energy Master Plan. As a part of this plan, the New Jersey Economic Development Authority and Board of Public Utilities has issued funding to assist in improving grid reliability in the state through CHP (New Jersey Clean Energy Program, Undated).

Other Programs

- **Combined Heat & Power Association (CHPA).** The CHPA is a membership organization that encourages increased deployment of CHP technologies. The

CHPA has worked with EPA and DOE to develop a number of resources to address existing barriers to the development of CHP technologies. Local governments can access many information resources through the CHPA website, including policies encouraging CHP, overviews of CHP basics, and several databases of CHP projects and resources.

Website: *http://chpassociation.org/*

- **Database of State Incentives for Renewable Energy (DSIRE).** A project of the U.S. Department of Energy, the North Carolina Solar Center, and the Interstate Renewable Energy Council, DSIRE provides information on federal, state, and local incentives for renewable energy and energy efficiency projects (including CHP), including tax credits, loans, and grants. The database also provides information on state and local regulations pertaining to renewable energy purchases and on-site renewable energy generation, including overviews of state and local net metering rules, renewable portfolio standards, and requirements for renewable energy use at public facilities.

Website: *http://www.dsireusa.org/*

- **dCHPP: CHP Policies and Incentives Database.** The dCHPP database is a project of the U.S. EPA Combined Heat and Power Partnership. The database provides users with a comprehensive collection of CHP policies and incentives, which can be searched for at the state or federal level. The database also includes a function for searching by policy/incentive type.

Website: *http://epa.gov/chp/policies/database.html*

9. CASE STUDIES

The following two case studies describe CHP projects implemented at local government facilities. Each case study describes how the project was initiated and key project activities, features, and benefits.

City of Boston, Massachusetts

The City of Boston adopted a Climate Action Plan in 2007, setting a goal of reducing GHG emissions 20 percent from current levels by 2020, and 80 percent by 2050 (Boston, 2007b). The city issued an update to this plan in 2011. The update gives credit to CHP systems as a major contributor to the reduction of GHG emissions in municipal operations (Boston, 2011). Since 2000, 11 CHP systems have been developed in Boston, two of which are located at public facilities. In 2013, the American Council for an Energy-Efficient Economy named Boston as the most energy efficient city in the United States (ACEEE, 2013).

PROGRAM INITIATION

Boston has a number of programs to encourage CHP development. The city's key policy driver is its climate action plan, which recognizes reductions in GHG emissions from CHP.

PROGRAM FEATURES

Boston has developed a number of CHP projects through its climate change and energy infrastructure improvement initiatives. Key activities include the following:

- **Climate Action Plan.** The city's 2007 plan describes a number of initiatives that could be used to promote CHP development. For example, it notes the Boston Housing Authority uses energy performance contracts (EPCs) that include CHP installations. EPCs can make the financing of energy efficiency projects more attractive, allowing the cost of the capital improvements to be paid for out of the savings generated by energy and water conservation measures.

- **Green Building Standards.** Boston is the first city in the United States to require a green building standard through municipal zoning requirements. The city requires projects larger than 50,000 square feet to meet the U.S. Green Building Council's LEED certification

standards. To comply with the green building directives, the Boston Housing Authority has adopted a "whole building" and environmentally conscious approach to its new projects. It is incorporating LEED standards and ENERGY STAR products into its projects, some of which include CHP units. To further encourage CHP, the city awards one credit point toward LEED certification for buildings drawing 10 percent of their total energy use from CHP systems.

- **Interconnection Standards.** The mayor's office joined the Distributed Generation (DG) Collaborative (now the DG Working Group), a forum hosted by the Massachusetts Technology Collaborative, to develop an interconnection tariff and to examine the role of DG in electricity distribution system planning. In 2012, the DG Working Group recommended changes to the uniform interconnection standards to simplify the process and costs for DG systems, including CHP. The Massachusetts Department of Public Utilities is reviewing the working group's recommendations (Massachusetts Distributed Generation Working Group, 2013).

- **District Heating.** Boston has one of the most extensive district heating systems in the country. The GenOn Kendall Station CHP system, a 277 MW natural gas-fired plant located in Cambridge, recently started providing steam to the City of Boston's district heating system. Boston is working on plans to expand its use of district heating and to incorporate CHP into these systems where possible (Boston Metro Green, 2011).

- **CHP in Hospitals.** The Boston Green Ribbon Commission, an organization composed of Boston business, institutional, and civic leaders in Boston working to develop climate change strategies, recently collaborated on a report evaluating hospital cost savings through the use of CHP. The report, *Powering the Future of Health Care – Financial and Operational Resilience: A Combined Heat and Power Guide for Massachusetts Hospital Decision Makers*, provides an overview of CHP applications, costs, and other requirements to help hospital executives decide whether to invest in CHP. The report found that installing a 1 MW CHP system at a hospital can result in a net positive cash flow of $700,000 annually and also provides emissions reduction and resilience benefits (Health Care Without Harm, 2013).

PROGRAM RESULTS

Between 2000 and 2012, the City of Boston assisted in installing a total of more than 2 MW of CHP at 11 facilities, including two office buildings, two multifamily buildings, a brewery, a hotel, an amusement park, a college, a jail, and two private residences. The neighboring city of Cambridge also installed eight CHP systems between 2000 and 2012. These CHP projects are achieving meaningful energy and cost savings and GHG reductions.

- Harvard University installed 75 kW CHP system at a hotel owned by the University to helps meet its GHG reduction goals. The CHP system supplies 10 to 15 percent of the hotel's electricity, and the thermal output is used for space heating, hotel laundry, and domestic hot water. The system will avoid an estimated 175 metric tons of CO_2 equivalent annually, equivalent to the annual energy use from 24 homes (Sustainability at Harvard, 2011).

- The Mass Bay Harpoon Brewery installed a 225 kW natural gas-fired reciprocating engine in 2012. This CHP system achieves 90 percent efficiency, provides 70 percent of the brewery's electric power, and supplies 90 percent of its hot water needs (Harpoon Brewery, 2012).

- Two private residences in Boston installed micro CHP systems in 2009. One of these, a 1.2 kW unit, helped a Boston homeowner save $2,000 in energy costs during a single heating season (November through March, 2009) (Ferguson, 2009).

Website: *http://www.cityofboston.gov/ environmentalandenergy/*

City of Hartford, Connecticut

Hartford, Connecticut, has around 73 MW of installed CHP capacity at 13 sites (ICF, 2013). The CHP systems are located at public and private facilities, and include two downtown district energy systems: the Capitol Area System and the Hartford Steam district energy systems. As of 2013, there were two more regional systems under development with a CHP component: the Hartford Steam fuel cell CHP system addition, and the Parkville Cluster CHP microgrid project.

PROGRAM INITIATION

State policies favorable to CHP have had significant influence on the number of CHP systems installed in the city. As the state capital of Connecticut, Hartford has been involved in implementing state programs offering CHP incentives, innovative financing options, pilot programs, and the state's comprehensive energy strategy. This involvement, combined with the current city charter, *One City, One Plan*, has created opportunities for Hartford to use CHP to promote livable and sustainable neighborhoods. Public-private partnerships have played an important role in many of city's CHP projects, and some projects have received state grant support, while others are privately owned.

PROGRAM FEATURES

Hartford has been fertile ground for CHP since the city established its first CHP system in the 1940s. The city's CHP use has increased steadily since the year 2000, due in part to policies and initiatives at the state level, including the following:

Connecticut adopted a Climate Change Plan in 2005. The plan includes two action areas specifically encouraging clean CHP. The Climate Change Plan also recommends CHP be included as a third class of power generation in the state's Renewable Portfolio Standard (Connecticut, 2005). In response, Connecticut added CHP as a Class III resource to the RPS, with targets that began in 2007.

In 2005, the state established an incentive program to address grid congestion, which occurs when actual or scheduled flows of electricity over a line or piece of equipment are constrained below desired levels. CHP systems in congested areas received an added bonus of $50/kW over the base incentive of $200-$500/kW. Eighty-one CHP projects, for a total installed capacity of 250 MW, received funding under this program until its closure in 2008 (USEA, 2011).

In 2008, Connecticut adopted Public Act No. 08-98, *An Act Concerning Global Warming Solutions*, which requires the state to reduce its GHG emissions to 10 percent below 1990 levels by 2020, and 80 percent below 2001 levels by 2050.

The state is a participant in the Regional Greenhouse Gas Initiative and has two CHP set-aside accounts for eligible CHP projects (Connecticut, 2008). (Note that in late 2013 the state proposed an amendment to eliminate one of its CHP set-asides and reduce the other.)

Connecticut approved Public Act No. 12-148, *An Act Enhancing Emergency Preparedness and Response*, in June 2012, which established the nation's first microgrid pilot program to support local distributed energy generation, including CHP for critical facilities (Connecticut, 2012).

Connecticut established a grant program providing up to $450 per kilowatt for CHP systems (5 MW or less) located in the service territories of Connecticut utilities (Energize Connecticut, 2013).

PROGRAM RESULTS

Between 2000 and 2012, the City of Hartford assisted in installing more than 4.78 MW of CHP capacity at 10 facilities, including two downtown district energy systems, three educational facilities, several private businesses, a zoo, a local Veterans Administration hospital, and a YMCA. The Hartford Steam Company installed a 7.5 MW CHP system in 1998 at Hartford Hospital, and is in the process of installing a 1.4 MW fuel cell CHP system adjacent to Hartford Hospital (IDEA, 2013). CHP systems in Hartford remained operational during Superstorm Sandy in 2012.

Using CHP, Hartford Steam reduced its downtown district heating system's water consumption by 41 percent between 2006 and 2009. Over the same period, the company's downtown system reduced air pollutants by 54 percent and greenhouse gas emissions by 23 percent (Hartford Business Journal, 2013). Compared with conventional combustion-based power generation, Hartford Steam's new fuel cell power plant is expected to avoid annual emissions of more than 57,000 pounds of nitrogen oxide, more than 128,000 pounds of sulfur dioxide, more than 3,000 pounds of particulate matter, and more than 6,000 metric tons of carbon dioxide—equivalent to the annual emissions from 1,200 cars (Hartford Steam, 2013).

Some of Hartford's CHP projects currently under development are part of a microgrid. Microgrids can provide a centralized energy supply (with CHP, the systems can provide both power and thermal energy needs) for multiple facilities located in close proximity. The City of Hartford's Parkville Cluster is a proposed microgrid project with CHP that has received a state grant of $2.06 million as part of a broader effort in Connecticut to encourage microgrid pilot programs (Connecticut, 2013). The proposed project will consist of a 600 kW gas turbine CHP system and will cover a school, senior center, and library, as well as an adjacent supermarket and gas station (Bourgeois, 2013).

Websites: *http://www.ct.gov/deep/cwp/view. asp?a=4120&Q=508780 and http://www.hartford.gov/*

10. ADDITIONAL EXAMPLES AND INFORMATION RESOURCES

Title/Description	Website
Examples of Combined Heat and Power Opportunities for Local Governments	
Schools	
Frankfort, Illinois. As part of additions to Lincolnway North High School in Frankfort, the school district added a CHP system to reduce electrical demand charges and save additional money. Electricity cost savings for the system were approximately $213,900 per year while saving 4,000 therms of energy yearly.	*http://www.berg-eng.com/ executive_team/hpacaug96.pdf*
Hartford, Connecticut. Designed to meet LEED Silver Certification, the Annie Fisher Magnet School in Hartford installed a CHP system among many other energy efficiency upgrades.	*http://online.qmags.com/LBD0411/ default.aspx?pg=100&mode=2*
Swampscott, Massachusetts. Swampscott High School installed a 75 kW CHP system that generated about 760 Massachusetts Alternative Energy Credits per year. Utility incentives brought the system payback period to only five years.	*http://www.groomenergy.com/case_ study_combined_heat_power.html*
Multi-Family Housing	
Bronx, New York. A large CHP system was installed at the Co-Op City Central Plant in 2010 in the Bronx, serving the largest co-op housing development in the country. The CHP system produces savings of approximately $13 million annually.	*http://www.combinedcyclejournal. com/5Q-Pacesetter/CoopCityCentral. pdf*
New Bedford, Massachusetts. The New Bedford Housing Authority installed a 75 kW CHP system at the Boa Vista Apartments, an elderly high-rise housing development. The system is expected to save the housing authority nearly $400,000 over 10 years, an overall total energy savings of 24 percent.	*http://www.aegisenergyservices. com/wp-content/uploads/2012/01/ BoaVistaApartments.pdf*
Watertown, Massachusetts. As part of a large two-year comprehensive energy and water conservation audit and improvement program, the Watertown Housing Authority installed a 60 kW CHP system for multiple building use. Collectively, the upgrades are projected to save the housing authority 883,976 kWh of electricity annually.	*http://www.reuters.com/ article/2008/10/01/idUS188401+01- Oct-2008+PRN20081001*
Wastewater Treatment Plants	
Auburn, New York. Auburn installed a 1400 kW CHP system at the local wastewater treatment plant in 2010. After the capital costs are repaid, savings from the system are expected to total $900,000.	*http://www.auburnny.gov/Public_ Documents/AuburnNY_PLanning/ cipproject*
Dallas, Texas. A 4,200 kW CHP system was installed at the Southside Wastewater Treatment Plant in Dallas. The system will allow energy to be sold to the city 3 cents/kWh cheaper than before it was installed. In addition, the CHP system will avoid an estimated 36,000 metric tons of CO_2 emissions.	*http://www.greendallas.net/pdfs/ greenTimes/newsletter_102009.pdf*

Title/Description	Website
Holt, Michigan. The Dehli Charter Township in Holt installed a 60 kW CHP system at the local wastewater treatment plant. The system, which went online in 2009, is expected to produce annual electricity savings of $30,000 and annual natural gas savings of $40,000.	*http://www.tpomag.com/editorial/2010/05/low-tech-high-tech*
Oceanside, California. A 560 kW CHP system installed at the San Luis Rey Wastewater Treatment Plant in Oceanside is capturing methane to produce electricity. As a result, the city is expecting to save $185,000 per year on electricity and $150,000 per year on natural gas.	*http://www.utsandiego.com/news/2007/dec/07/cogeneration-sewage-plan-will-save-city-335000-uti/?print&page=all*
Toledo, Ohio. A 10 MW system was installed at the Bay View Wastewater Treatment plant in Toledo in 2010, the largest wastewater treatment facility in Northwest Ohio. The system is expected to avoid more than 650,000 metric tons of CO_2 over the lifetime of the system.	*http://www.epa.gov/lmop/documents/pdfs/conf/14th/ellman.pdf*
Vineland, New Jersey. A 170 kW CHP system was installed at the Landis Sewerage Authority Wastewater Treatment plant in Vineland in 2008, the largest facility in New Jersey. The project has a total value of more than $1 million.	*http://www.reedconstructiondata.com/building-types/water-treatment/new-jersey/projects/1000606517/*
Municipal Utilities	
Austin, Texas. In 2004, Austin Energy installed a 4 MW CHP system at its Domain Industrial Park. The energy produced from the system is used to provide district cooling services, reducing operational and energy costs while enhancing commercial property value.	*http://www.austinenergy.com/commercial/other%20services/On-Site%20Energy%20Systems/districtcooling.htm*
Gainesville, Florida. In 2009, Gainesville Regional Utilities partnered with Shandis HealthCare to construct the GRU South Energy Center, a 4.3 MW CHP plant that will provide 100 percent of the energy needs at the University of Florida's Shandis Cancer Hospital. The projected annual energy savings is equivalent to the power needed to run more than 3,000 homes.	*http://www.burnsmcd.com/Press-Releases/Detail/GRU-Shands-HealthCare-partner-on-unique-energy-project*
Rochester, New York. The County of Monroe had a CHP system installed at the Rochester International Airport in 2002. The system will produce $500,000 in savings annually, reducing the airport's energy consumption by 47 percent. This is the equivalent of the amount of jet fuel two 737's would use flying from Rochester to Chicago for a whole year.	*https://www.nyserda.ny.gov/Energy-Efficiency-and-Renewable-Programs/CHP/CHP-Conferences/-/media/Files/EIBD/Research/Combined%20Heat%20and%20Power/1GRIALynchSlaybaugh.ashx*
Landfill Gas Energy	
Albany, Georgia. A 1.9 MW CHP system was installed at the Dougherty County Landfill in Albany in 2011. The energy produced is sold to the nearby Marine Corps Logistic Base. In combination with other installed energy efficiency measures at the base, the CHP system will reduce the base's GHG emissions by 9,300 metric tons annually, equivalent to the annual energy use of more than 1,200 homes.	*http://www.epa.gov/chp/partnership/current_winners.html#six*
Kalispell, Montana. A CHP plant is currently in the process of being constructed in Kalispell in Flathead County. The system will use wood byproducts from a local lumber company's logging and mill operations to generate about 2.5 MW of power, enough to qualify the project for renewable energy credits.	*http://montananewsnow.com/the-latest/tag/fh-stoltze-co-generation-plant*
La Crosse, Wisconsin. Energy created from the La Crosse County Landfill CHP system is transported through a two-mile-long underground pipe to provide green power to the local grid and heat both buildings and water on the Hendersen Health System's campus.	*http://www.epa.gov/lmop/partners/award/2012.html*
Midland, Michigan. In 2011, the City of Midland had a CHP system constructed at the city landfill. The system will increase plant efficiency and add a revenue stream. The first of the two CHP generators already constructed saved the facility an estimated $400,000 in energy costs in the first year of operation. When the second generator comes online, the facility expects to generate an excess of energy that would then be sold back to a local utility.	*http://www.mlive.com/midland/index.ssf/2011/01/trash_to_treasure_midland_landfill_converting_methane_into_electricity_on_schedule_to_fire_up_next_m.html*

Title/Description	Website
Punta Gorda, Florida. In 2011, an energy developer constructed a 4.2 MW CHP plant at the Zemel Road Landfill in Punta Gorda. The project is expected to generate long-term recurring revenue and earnings, and a strong return on investment for company stockholders.	*http://www.lime-energy.com/about/ news/Lime-Energy-Announces-Acquisition-of-Landfill-Gas-Rights-for-Development-of-42MW-Electricity-Generating-Facility*
Trinity, Alabama. Part of the energy created by the CHP systems at the Morgan County Landfill in Trinity provides winter heating for the city's newly constructed recycling center. EPA recognized this achievement by naming the project the 2011 Landfill Methane Outreach Program Community Partner of the Year.	*http://www.epa.gov/lmop/partners/ award/2011.html*
District Energy Systems	
Philadelphia, Pennsylvania. A 200 kW CHP microturbine was installed at Philadelphia Gas Work's headquarters in 2011. The CHP system is expected to provide an 84 percent reduction in NO_x, 100 percent reduction in SO_2, and 33 percent in CO_2, or 475 metric tons of avoided carbon dioxide emissions per year. This is equivalent to the annual emissions of 87 cars. The project is also expected to save the company $130,000 per year.	*http://www.winningcombo. net/succeed/ PGWinstallsCHPmicroturbine.pdf*
Tucson, Arizona. NRG Thermal owns and operates a CHP plant in Tucson. The system reduces both energy costs and emissions while enhancing power reliability.	*http://www.nrgthermal.com/chp.htm*
City/Local Climate Change Plans	
Berkeley, California. The City of Berkeley adopted its Climate Action Plan in 2009. The plan outlines a series of recommendations the local government can take to reduce Berkeley's GHG emissions by 80 percent by 2050. One of the recommendations in the plan is to "investigate the potential and possible sites for combined heat and power (CHP) systems in Berkeley."	*http://www.cityofberkeley.info/ uploadedFiles/Planning_and_ Development/Level_3_-_Energy_ and_Sustainable_Development/ Berkeley%20Climate%20Action%20 Plan.pdf*
Chicago, Illinois. Chicago's Climate Action Plan 2008 outlines 35 actions that can be taken to reduce GHG emissions within the city. The plan proposes a goal to achieve GHG emissions reduction of 25 percent below 1990 levels by 2020 and 80 percent below 1990 levels by 2050. The report's includes an action area to increase efficient power generated on-site using DG and CHP systems.	*http://www.chicagoclimateaction. org/filebin/pdf/finalreport/ CCAPREPORTFINALv2.pdf*
Louisville, Kentucky. In 2009, the Louisville Metro Government (LMG) released a series of recommendations for reducing the city's GHG emissions. Recommendation 58 states: "LMG should investigate and work to remove barriers and provide incentives to stimulate greater adoption of combined heat and power systems (CHP). This includes issues of appropriate environmental regulations, utility interconnection policies, utility tariffs and reasonable financial incentives for high performance CHP systems."	*http://www.louisvilleky.gov/ NR/rdonlyres/4A0D4B18-885B-4A48-A803-A7A25EA1688E/0/ FinalClimateActionReport.pdf*
New York, New York. PlaNYC's April 2011 update report details the challenges New York City faces in the coming years and decades in terms of climate change, and outlines recommendations and initiatives to meet those challenges. Initiative 13 encourages the development of clean distributed generation and CHP systems, and mentions several examples of CHP development sites in the city.	*http://nytelecom.vo.llnwd.net/ o15/agencies/planyc2030/pdf/ planyc_2011_planyc_full_report.pdf*
Philadelphia, Pennsylvania. In 2007, Philadelphia released a series of recommendations aiming to reduce the city's GHG emissions. Recommendation 8 states: "Promote the installation of combined heat and power systems (cogeneration) at City complexes. (CG) Combined heat and power systems can reduce GHG emissions through their high fuel efficiencies, and these systems may be cost effective at some City-owned facilities and at large private-sector projects. Initially, the City will evaluate the financial feasibility of cogeneration at the Prison System's complex located on State Road. *Going forward, the City will explore approaches to encourage combined heat and power systems at other public and private facilities.*"	*http://www.phila.gov/green/PDFs/ Attachment1_Philadelphia_Local_ Action_Plan_Climate_Change.pdf*

Title/Description	Website
Information Resources for Combined Heat and Power Opportunities for Local Governments	
Overview	
Combined Heat and Power: Frequently Asked Questions. This EPA CHP partnership document answers several common questions such as how CHP works, what facilities use CHP, and the benefits and costs of CHP.	*http://www.epa.gov/chp/documents/ faq.pdf*
Catalog of CHP Technologies. This EPA CHP partnership document provides an overview of how CHP systems work, the key concepts of efficiency and power-to-heat ratios, and summarizes the cost and performance characteristics of commercially proven CHP technologies.	*http://www. cleanenergyresourceteams.org/ sites/default/files/publication_files/ CERTsManualCh10.pdf*
Biomass Combined Heat and Power Catalog of Technologies. This EPA CHP partnership document provides a detailed technology characterization of biomass CHP systems. The report reviews the technical and economic characterization of biomass resources, biomass preparation, energy conversion technologies, power production systems, and complete integrated systems.	*http://www.epa.gov/chp/documents/ biomass_chp_catalog.pdf*
Waste Heat to Power Systems. This EPA CHP partnership document examines the recovery of industrial waste heat for power (WHP), a largely untapped type of CHP. The report explains the opportunity for WHP, applicable technologies, industrial WHP applications, and the economics and market status of WHP.	*http://www.epa.gov/chp/documents/ waste_heat_power.pdf*
Distributed Energy Resources Guide. This California Energy Commission guide provides information on performance, costs, strengths, weaknesses, and future development of CHP systems.	*http://www.energy.ca.gov/distgen/ equipment/chp/chp.html*
Benefits of Combined Heat & Power	
Combined Heat and Power: Effective Energy Solutions for a Sustainable Future. This Oak Ridge National Lab report describes four key areas in which CHP is effective and holds promise for the future: as an environmental solution, a competitive business solution, a local energy solution, and an infrastructure modernization solution. The appendix provides basic information about CHP.	*http://info.ornl.gov/sites/ publications/files/Pub13655.pdf*
Benefits of CHP. This EPA CHP Partnership web page provides information on the efficiency, reliability, environmental, and economic benefits CHP provides.	*http://www.epa.gov/chp/basic/index. html*
CHP Emissions Calculator. EPA has developed this tool to assist CHP project developers and policy makers in estimating the environmental benefits of installing CHP systems. The calculator allows users to characterize a model CHP system and compare its benefits with a comparable separate heat and power system.	*http://www.epa.gov/chp/basic/ calculator.html*
CHP Calculation Methodology for LEED-NC v2.2 EA Credit 1. This document provides guidance on accounting for CHP systems when using the U.S. Green Building Council's Leadership in Energy and Environmental Design (LEED) v2.2 rating system for new construction.	*http://www.utexas.edu/utilities/ sustainability/leed/documents/ CHPCalculationMethodology.pdf*
Treatment of District or Campus Thermal Energy in LEED V2 and LEED 2009 – Design & Construction. This document describes the treatment of district and campus thermal energy in the LEED v2.x and LEED-2009 Design & Construction and Interior Design & Construction rating systems.	*http://www.usgbc.org/Docs/Archive/ General/Docs7671.pdf*
Opportunities for Combined Heat & Power	
Municipal Wastewater Treatment Facilities. This EPA CHP website provides information on the compatibility of CHP systems with wastewater treatment facilities.	*http://www.epa.gov/CHP/markets/ wastewater.html*
Landfill Gas As A Fuel for Combined Heat and Power. This paper describes opportunities to capture LFG from landfills and use it as a source of fuel for CHP systems in various applications.	*http://www.energyvortex.com/ files/Landfill_Gas_as_Fuel_for_ Combined_Heat_and_Power.pdf*

Title/Description	Website
The Role of District Energy in Greening Existing Neighborhoods. This report, from the Preservation Green Lab, looks at how district energy can be a critical element of a successful community energy plan for neighborhoods. It describes what district energy is, why it matters, how to develop district energy systems, and case studies from around North America illustrating the crucial role of city governments in promoting and implementing district energy.	*http://newenergycities.org/ resources/greening-existing- neighborhoods-a-district-energy- policy-primer/view*
Community Energy: Planning, Development & Delivery. An International District Energy Association guidebook that provides an overview of the local energy project development process to assist mayors, planners, community leaders, real estate developers and economic development officials in making informed decisions on the analysis, planning, development and delivery of district energy systems.	*http://districtenergy.org/community- energy-planning-development-and- delivery/*
Combined Heat and Power: Enabling Resilient Energy Infrastructure for Critical Facilities. This report summarizes how critical infrastructure facilities with CHP systems were able to power through Superstorm Sandy. Several examples from other storms and blackout events in other regions of the country are also included. This report also provides information on the use of CHP for reliability purposes, as well as state and local policies designed to promote CHP in critical infrastructure applications.	*https://www1.eere.energy.gov/ manufacturing/distributedenergy/ pdfs/chp_critical_facilities.pdf*
Department of Housing and Urban Development (HUD) Resources on CHP. This website describes HUD energy initiatives, policies, and how federal government-wide energy policies affect HUD programs and assistance. The portal includes resources developed by HUD to explain CHP to building owners and managers, and provides evaluation tools specific to multifamily housing.	*http://portal.hud.gov/hudportal/ HUD?src=/program_offices/comm_ planning/library/energyevelopment/ Guide1QAMultifamilyHousing.pdf*
Market Analyses. These analyses cover a wide range of markets, including commercial and institutional buildings and facilities, district energy, and industrial sites. They also examine the market potential for CHP at federal sites and in selected states and regions.	*https://www1.eere.energy.gov/ manufacturing/distributedenergy/ market_analyses.html*
Combined Heat and Power on Brownfield Sites. This report, from Redevelopment Economics, organizes the federal and New York State energy incentives available for CHP and explores federal policy issues surrounding CHP, district energy, and brownfields. It also discusses and analyzes a number of in-depth case studies, including financing, technology, impacts, and how CHP fits into the overall redevelopment project.	*http://redevelopmenteconomics. com/yahoo_site_admin/assets/ docs/Brownfields-CHP-district_ Final.35291029.pdf*
Key Participants	
Portfolio Standards and the Promotion of Combined Heat and Power. This EPA CHP Partnership paper discusses the different ways CHP is incorporated in portfolio standards.	*http://epa.gov/chp/documents/ ps_paper.pdf*
Combined Heat and Power: A Clean Energy Solution. This report by DOE and EPA examines the benefits of CHP, the current status of CHP, its potential and future role in the United States, drivers and barriers to CHP deployment, and policy solutions to promote CHP.	*http://www1.eere.energy.gov/ manufacturing/distributedenergy/ pdfs/chp_clean_energy_solution.pdf*
Combined Heat and Power: A Resource Guide for State Energy Officials. This NASEO resource guide provides a technology and market overview of CHP and ways in which state officials can support CHP through energy assurance planning, energy policies and utility regulations, and funding/financing opportunities for CHP.	*http://www.naseo.org/data/sites/1/ documents/publications/CHP-for- State-Energy-Officials.pdf*
Challenges Facing Combined Heat & Power Today: A State-by-State Assessment. This ACEEE report examines the CHP environment and barriers to CHP development both in general and then by state. The report concludes with suggestions on how stakeholders can further the CHP market building on existing successes.	*http://aceee.org/research-report/ ie111*
Guide to the Successful Implementation of State Combined Heat and Power Policies. This guide, from the State & Local Energy Efficiency Action Network, provides state utility regulators and other state policymakers with actionable information to assist them in implementing key state policies that affect CHP.	*http://www1.eere.energy.gov/ seeaction/chp_policies_guide.html*

Title/Description	Website
Project Development	
CHP Project Development Handbook. This EPA CHP Partnership handbook provides information, tools, and insights on the project development process, CHP technologies, and the resources of the CHP Partnership.	*http://www.epa.gov/chp/documents/chp_handbook.pdf*
Spark Spread Estimator. This EPA CHP Partnership resource calculates the difference between the delivered electricity price and the total cost to generate power with a prospective CHP system, helping to easily evaluate a prospective CHP system for its potential economic feasibility.	*http://www.epa.gov/chp/project-development/stage1.html*
Costs and Funding Opportunities	
dCHPP database. EPA's dCHPP is an online database that allows users to search for CHP policies and incentives at the local, state and federal level.	*http://www.epa.gov/chp/policies/database.html*
Database of State Incentives for Renewable Energy. DSIRE is a comprehensive source of information on incentives and policies that support renewables and energy efficiency, including CHP, in the United States.	*http://www.dsireusa.org/incentives/index.cfm?EE=1&RE=1&SPV=0&ST=0&technology=combined_heat_power&sh=1*
Case Studies & Strategies for Effective Project Implementation	
Distributed Energy Case Study Database. DOE maintains this database of CHP projects. Users can narrow database searches based on state, Clean Energy Application Center, market sector, NAICS code, system size, technology, fuel, thermal energy use, and year installed.	*http://www1.eere.energy.gov/manufacturing/distributedenergy/chp_projects.html*
Combined Heat and Power Installation Database. This database, operated by ICF International, lists all known CHP installations. It also lists and provides information on the CHP units, organized by state.	*http://www.eea-inc.com/chpdata/*
Combined Heat and Power Systems: Improving the Energy Efficiency of Our Manufacturing Plants, Buildings, and Other Facilities. This Natural Resources Defense Council report provides 30 case studies demonstrating how various industrial and manufacturing facilities have benefited from using CHP.	*http://www.nrdc.org/energy/files/combined-heat-power-IP.pdf*
District Energy Case Studies. This International District Energy Association web page provides multiple district energy case studies from across the United States and around the world.	*http://www.districtenergy.org/case-studies*
Regional Studies	
Midwest Manufacturing Snapshot: Energy Use & Efficiency Policies. This World Resources Institute paper offers a snapshot of industrial energy use and current state approaches to reducing industrial energy intensity and energy costs for manufacturers. It also provides state-by-state policy studies for 10 Midwestern states.	*http://pdf.wri.org/working_papers/midwest_manufacturing_snapshot/midwest_manufacturing_snapshot.pdf*
Power Almanac of the American Midwest. This World Resources Institute interactive map provides facts and figures about the use and potential of CHP in the Midwest.	*http://www.wri.org/project/midwest-almanac#map:stt=mw&res=chp&gas=all*
Clean Energy Roadmap: Washington State. This report by the Cascade Power Group outlines three scenarios to help meet Washington State's energy demands and emissions reduction goals by 2035 while decreasing total energy consumption, including the employment of CHP systems. It is also designed to complement the state's 2012 State Energy Strategy.	*http://www.northwestcleanenergy.org/NwChpDocs/WA%20Clean%20Energy%20Roadmap%202012.pdf*
California SGIP & CHP: Recent History and Current Status of the California Self-Generation Incentive Program. This paper highlights the key changes and new program rules for the 2011-2014 California Self-Generation Incentive Program. The paper is intended to assist potential CHP site owners, project developers, and others with understanding the latest SGIP program rules.	*http://www.pacificcleanenergy.org/RESOURCES/Library/PDF/SGIP2011Fullpaper-FINAL.pdf*

Title/Description	Website
2008 Combined Heat and Power Baseline Assessment and Action Plan for the California Market. This report provides an updated baseline assessment and action plan for combined heat and power in California and identifies the hurdles preventing the expanded use of CHP systems.	*http://www.pacificcleanenergy.org/ RESOURCES/Library/PDF/PRAC_CA_ Plan_2008.pdf*
2011 Combined Heat and Power and Other Clean Energy System Baseline Assessment and Action Plan for the Nevada Market. This report assesses and summarizes the current status of combined heat and power, district energy, and waste heat-to-power in Nevada, and identifies the hurdles preventing the expanded use of these clean energy systems.	*http://www.pacificcleanenergy.org/ RESOURCES/Library/PDF/PCEAC_ NV_Plan_2011.pdf*
2011 Combined Heat and Power and Other Clean Energy System Baseline Assessment and Action Plan for the Hawaii Market. This report assesses and summarizes the current status of combined heat and power, district energy, and waste heat-to-power in Hawaii, and identifies the hurdles preventing the expanded use of these clean energy systems.	*http://www.pacificcleanenergy.org/ RESOURCES/Library/PDF/PCEAC_HI_ Plan_2011.pdf*
Combined Heat and Power in Texas: Status, Potential, and Policies to Foster Investment. This study from Summit Blue Consulting examines CHP installations in Texas; assesses the technical, economic, and regulatory environment surrounding CHP development; and identifies policy options to encourage greater investment in CHP.	*http://www.gulfcoastcleanenergy. org/Portals/24/Reports_studies/ Summit%20Blue%20CHP%20 Study%20to%20PUCT%20081210.pdf*
U.S. DOE CHP Technical Assistance Partnerships. DOE's seven CHP Technical Assistance Partnerships, formerly called the Clean Energy Application Centers, promote and assist in transforming the market for CHP, waste heat to power, and district energy technologies and concepts throughout the United States.	*http://www1.eere.energy.gov/ manufacturing/distributedenergy/ chptaps.html*

11.   REFERENCES

Ameresco. 2011. *Southeastern Regional School District.* Available: http://www.ameresco.com/sites/default/files/southeasternsd_0.pdf. Accessed 12/13/2013.

American Council for an Energy-Efficient Economy (ACEEE). 2011. *Sheboygan Wastewater Treatment Plant Energy Efficiency Initiatives.* Available: http://aceee.org/sector/local-policy/case-studies/sheboygan-wastewater-treatment-plant-. Accessed 12/13/2013.

American Council for an Energy-Efficient Economy (ACEEE). 2013. *Report Ranks U.S. Cities' Efforts to Save Energy.* Available: http://www.aceee.org/press/2013/09/report-ranks-us-cities-efforts-save-. Accessed 12/13/2013.

Baltimore City Department of Public Works. 2012. *Cogeneration Facility.* Available: http://publicworks.baltimorecity.gov/Bureaus/WaterWastewater/Wastewater/BackRiverWastewaterTreatmentPlant/CogenerationFacility.aspx. Accessed 8/29/2013.

BioCycle. 2012. *Microturbines Fill Biogas Utilization Niche.* Available: http://www.biocycle.net/2012/08/14/microturbines-fill-biogas-utilization-niche/. Accessed 12/13/2013.

Boston. 2007a. *Article 37: Green Buildings.* Available: http://www.masstech.org/renewableenergy/public_policy/DG/resources/2006-Boston-Zoning-Article37-ModernGrid.pdf. Accessed 8/28/2007.

Boston. 2007b. *The City of Boston's Climate Action Plan, December 2007.* Available: http://www.cityofboston.gov/climate/pdfs/CAPJan08.pdf. Accessed 12/13/2013.

Boston. 2011. *A Climate of Progress: City of Boston Climate Action Plan Update 2011.* Available: http://www.cityofboston.gov/Images_Documents/A%20Climate%20of%20Progress%20-%20CAP%20Update%202011_tcm3-25020.pdf. Accessed 8/12/2013.

Boston Metro Green. 2011. *The Boston District Heating System: A Discussion with Jim Hunt, Chief of Environment and Energy for the City of Boston and Bill DiCroce, Chief Operating Officer of Veolia Energy.* Available: http://bostongreen.wordpress.com/2011/11/07/the-boston-district-heating-system-a-discussion-with-jim-hunt-chief-of-environment-and-energy-for-the-city-of-boston-and-bill-dicroce-chief-operating-officer-of-veolia-energy/. Accessed 12/13/2013.

Bourgeois, Tom. 2013. *Combined Heat and Power in Critical Infrastructure Applications.* Available: http://chpassociation.org/wp-content/uploads/2013/05/CHPA-Spring-Forum-PPTs-pt2.pdf. Accessed 12/13/2013.

Cambridge Public Schools. Undated. *Cambridge Rindge and Latin School Saves Energy with Help from NSTAR.* Available: http://www3.cpsd.us/schools/crls_nstar. Accessed 12/13/2013.

CEC. 2012. *Combined Heat and Power: Policy Analysis and 2011-2030 Market Assessment.* Available: http://www.energy.ca.gov/2012publications/CEC-200-2012-002/CEC-200-2012-002.pdf. Accessed 8/29/2013.

Chevron. 2006. *Case Study: City of Millbrae, California.* Available: http://www.ci.millbrae.ca.us/pdf/co-genprojectcasestudy.pdf. Accessed 5/20/2008.

Chicago. 2001. *Chicago Energy Plan.* Available: http://www.ci.chi.il.us/webportal/COCWebPortal/COC_EDITORIAL/2001EnergyPlan.pdf. Accessed 5/23/2008.

Chicago. 2007. *Assessment of CHP Goals for the Chicago Energy Plan: 2000-2005.* Available: http://egov.cityofchicago.org/webportal/COCWebPortal/COC_EDITORIAL/AssessmentofCHPGoals.pdf. Accessed 5/20/2008.

Chicago. 2008. *Cogeneration.* Available: http://egov.cityofchicago.org/city/webportal/portalContentItemAction.do?contentOID = 536912225&contentTypeName = COC_EDITORIAL&topChannelName = Dept&channelId = 0&programId = 0&entityName = Environment&deptMainCategoryOID = -536887205. Accessed 4/8/2008.

Cogeneration and On-Site Power Production. 2011. *U.S. Treatment Plant Converts High-Strength Waste to Energy.* Available: http://www.cospp.com/articles/print/volume-12/issue-3/project-profiles/us-treatment-plant-converts-high-strength-waste-to-energy.html. Accessed 10/25/2013.

Combined Cycle Journal. 2010. *Cogen Plant Reduces Energy Cost, Emissions, While Improving Service Reliability.* Available: http://combinedcyclejournal.com/3Q2010/31074-79Orange.w.pdf. Accessed 10/28/13.

Connecticut. 2005. *Climate Change Action Plan.* Available: http://www.ct.gov/deep/lib/deep/climatechange/ct_climate_change_action_plan_2005.pdf. Accessed 12/13/2013.

Connecticut. 2008. *Control of Carbon Dioxide Emissions.* Available: http://www.ct.gov/deep/lib/deep/air/regulations/mainregs/22a-174-31.pdf. Accessed 12/13/2013.

Connecticut. 2012. *An Act Enhancing Emergency Preparedness and Response.* Available: http://www.cga.ct.gov/2012/act/pa/pdf/2012PA-00148-R00SB-00023-PA.pdf. Accessed 12/13/2013.

Connecticut. 2013. *Governor Malloy Announces Nation's First Statewide Microgrid Pilot.* Available: http://www.governor.ct.gov/malloy/cwp/view.asp?A=4010&Q=528770. Accessed 12/13/2013.

District Energy. 2008. *Combined Heat and Power.* Available: http://www.districtenergy.com/CurrentActivities/chp.html. Accessed 5/24/2008.

EIA. 2013. *Average Retail Price of Electricity to Ultimate Customer: Total by End-Use Sector.* Available: http://www.eia.gov/electricity/data.cfm#sales. Accessed 11/20/09.

EBMUD. 2001. *East Bay Municipal Utility District Success Story.* Available: http://www.energy.ca.gov/process/pubs/ebmud.pdf. Accessed 5/20/2008.

EBMUD. 2006. *Project Profile: East Bay Municipal Utility District.* Available: http://www.chpcentermw.org/rac_profiles/pacific/EBMUD_v1_2.pdf. Accessed 3/17/2008.

Energize Connecticut. 2013. *Combined Heat and Power Pilot Program.* Available: http://www.energizect. com/businesses/programs/Combined-Heat-Power. Accessed 12/13/2013.

Energy Trust. 2006. *Biopower: City of Gresham Wastewater Services.* Available: http://files.harc.edu/Sites/ GulfCoastCHP/CaseStudies/GreshamORWastewaterServices.pdf. Accessed 3/17/2008.

Energy Vortex. Undated. *Vermont Wastewater Treatment Facility to Generate its Own Heat and Power.* Available: http://www.energyvortex.com/pages/ headlinedetails.cfm?id=682&archive=1. Accessed 10/28/2013.

Epping. 2007. *Article 22: Energy Efficiency and Sustainable Design.* Available: http://nhplanning.com/epping/ Article22/ARTICLE22.pdf. Accessed 8/28/2007.

Ferguson, Kevin. 2009. *A Winter's Tale: My First Season with Micro-Combined Heat and Power.* Available: http://green.blogs.nytimes.com/2009/04/29/a-winters-tale-my-first-season-with-micro-combined-heat-and-power/?_r=0. Accessed 12/13/2013.

GRU. 2010. *Using CHP to Enhance Energy Security. GRU South Energy Center at Shands Cancer Hospital.* Available: http://www.chpcon2011.com/Portals/24/Events/ CHP_trade_show_2010/Presentations/Heidt_Gainesville.pdf. Accessed 8/29/2013.

Harpoon Brewery. 2012. *Upgrade: Our New Cogeneration Unit!* Available: http://www.harpoonbrewery. com/blog/520/Upgrade-Our-new-Cogeneration-unit. Accessed 12/13/2013.

Hartford Business Journal. 2013. *Steam Power: Underground System of Tunnels and Piping Heats and Cools Downtown Hartford.* Available: http://www. hartfordbusiness.com/article/20130826/PRINTEDITION/308229928. Accessed 12/13/2013.

Hartford Steam. 2013. *Hartford Steam to Tap Fuel Cell Benefits.* Available: http://www.hartfordsteam.com/ news/intheloop.2nd.2013.pdf. Accessed 12/13/2013.

Health Care Without Harm. 2013. *Powering the Future of Health Care—Financial and Operational Resilience: A Combined Heat and Power Guide for Massachusetts Hospital Decision Makers.* Available: http://www.greenribboncommission.org/downloads/ CHP_Guide_091013.pdf. Accessed 12/13/2013.

ICF. 2013. *CHP Installation Database.* Maintained for the Department of Energy by ICF International. Available: http://www.eea-inc.com/chpdata/index. html. Accessed 8/29/2013.

IDEA. 2007. *Personal Communication with Rob Thornton, President. 8/7/2008. IDEA. 2013. What is District Energy?* Available: http://www.districtenergy.org/ what-is-district-energy. Accessed 8/29/2013.

IEA. 2013. *District Heating and Cooling: Environmental Technology for the 21st Century.* Available: http://iea-dhc.org/home.html. Accessed 8/29/2013.

IDEA. 2013. *FuelCell Energy announces MW-class power plant order from Hartford Steam for hospital in Connecticut.* Available: http://www.districtenergy. org/blog/2013/04/18/fuelcell-energy-announces-mw-class-power-plant-order-from-hartford-steam-for-hospital-in-connecticut/. Accessed 12/13/2013.

Lansing. 2013. *State of the City Address by Mayor Virg Bernero.* Available: http://www.lansingmi.gov/media/ view/2013_SOC_Speech_FINAL/3598. Accessed 12/13/2013.

Lawrence Berkeley Laboratory. 2010. *Energy Efficiency Services Sector: Workforce Size and Expectations for Growth.* Available: http://eetd.lbl.gov/ea/emp/reports/ lbnl-3987e.pdf. Accessed 9/20/12.

Maine Public Utilities Commission. Undated. *Maine Green Power Program: Frequently Asked Questions.* Available: http://www.maine.gov/mpuc/greenpower/ faq.shtml. Accessed 12/13/2013.

Massachusetts Distributed Generation Working Group. 2013. Available: http://massdg.raabassociates. org/. Accessed 12/13/2013.

Millbrae. 2005. *City Report, August 2005.* Available: http://www.ci.millbrae.ca.us/pdf/august05millbrae-newsletter.pdf. Accessed 5/20/2008.

MLive. 2013. *Lansing Board of Water & Light's New $182 Million REO Town Cogeneration Plant Goes Online.* Available: http://www.mlive.com/lansing-news/index.ssf/2013/07/lansing_board_of_water_lights.html. Accessed 12/13/2013.

Modesto. 2006. *Approval of Agreement to Carollo Engineers for Consulting Services.* Available: http://www.ci.modesto.ca.us/ccl/agenda/ar/2006/06/ar060613-14.pdf. Accessed 8/1/2008

MTC. 2007. *MA DG Collaborative.* Available: http://dg.raabassociates.org/. Accessed 8/28/2007.

NECHP. 2006. *Promoting Energy Efficiency, Environmental Protection, and Jobs through CHP.* Available: http://masstech.org/renewableenergy/public_policy/DG/resources/2006-12-NECHPI-MA-Energy-Environment-Transition.pdf. Accessed 5/20/2008.

New Jersey Clean Energy Program. Undated. *Combined Heat and Power and Fuel Cells.* Available: http://www.njcleanenergy.com/CHP. Accessed 12/13/2013.

NYSERDA. 2008. *NYSERDA CHP Program.* Available: http://www.nyserda.org/programs/dgchp.asp. Accessed 5/20/2008.

NYSERDA. 2013. *Improving Energy Resilience of Buildings in New York City.* Available: http://www.earth.columbia.edu/sitefiles/file/education/capstone/spring2013/Improving-Energy-Resilience-buildings-nyc.pdf. Accessed 8/29/2013.

Oak Ridge National Laboratory (ORNL). 2013. *Combined Heat and Power: Enabling Resilient Energy Infrastructure for Critical Facilities.* Available: https://www1.eere.energy.gov/manufacturing/distributedenergy/pdfs/chp_critical_facilities.pdf. Accessed 8/29/2013.

Philadelphia. 2007. *Local Action Plan for Climate Change.* Available: http://dvgbc.org/sites/default/files/resources/PhiladelphiaClimateChangeLocalActionPlan2007.pdf. Accessed 12/13/2013.

Philadelphia Inquirer. 2011. *Microturbines can Save on Energy, PGW Says.* Available: http://articles.philly.com/2011-10-16/business/30286400_1_natural-gas-microturbines-heat-and-power. Accessed 12/13/2013.

PPL Renewable Energy. 2012. *Lycoming County Landfill.* Available: http://www.lyco.org/Portals/1/ResourceManagementServices/Documents/ppl%20cogen.pdf. Accessed 12/13/2013.

PPM Energy. 2007. *About Klamath Cogeneration Plant.* Available: http://www.ppmenergy.com/klamath.html. Accessed 8/27/2007.

Renewable Energy from Waste. 2013. *Going All the Whey.* Available: http://www.rewmag.com/rew0413-kraft-foods-electricity-conversion.aspx. Accessed 8/29/2013.

RMT. 2008. *School Benefits from Landfill Gas.* Available: http://www.rmtinc.com/public/Awards/165.pdf. Accessed 4/8/2008.

San Francisco. 2007. *An Assessment of Cogeneration for the City of San Francisco.* Available: http://www.sfenvironment.org/downloads/library/ciscocogenerationreportpdf.pdf. Accessed 4/8/2008.

San Jose. 2008. *Instructions for Special Use Permit for Standby or Backup Power Generation and Co-Generation Facilities.* Available: http://www.sanjoseca.gov/planning/applications/dev_sup_gen_app.pdf. Accessed 4/8/2008.

Santa Cruz. 2003. *Wastewater Treatment Facility Programs.* Available: http://www.ci.santa-cruz.ca.us/pw/wastewt/cogeneration.html. Accessed 5/19/2008.

Santa Monica. 2003. *City of Santa Monica Solar Energy Resource, Co-Generation, and Energy Efficiency Program Development: Potential Study.* Available: http://www.smgov.net/cityclerk/council/agendas/2006/20060314/s2006031408-B-1.htm. Accessed 5/20/2008.

Shanahan, Mark, and Patrick D'Addario. 2013. *Challenges and Pathways to Deployment of CHP at Wastewater Treatment Facilities in Ohio.* Available: http://aceee.org/files/proceedings/2013/data/papers/2_221.pdf. Accessed 12/13/2013.

Skanska. Undated. *State of California, DGS Central Plant.* Available: http://www.usa.skanska.com/Projects/Project/?pid=909&plang=en-us. Accessed 12/13/2013.

Solar Turbines. 2011a. *Calabasas Landfill Gas to Energy.* Available: https://mysolar.cat.com/cda/files/2778631/7/dslfg-cl.pdf. Accessed 8/29/2013.

Solar Turbines. 2011b. *Bay View Wastewater Treatment Plant.* Available: http://mysolar.cat.com/cda/files/2864587/7/dschp-bvwtp.pdf. Accessed 12/13/2013.

Sonoma County. 2011. *Sonoma County Dedicates New Ultra Clean Energy Plant.* Available: http://press.sonoma-county.org/content.aspx?sid=1018&id=1620. Accessed 12/13/2013.

St. Paul. 2013. *District Energy St. Paul.* Available: http://www.districtenergy.com/. Accessed 8/29/2013.

Sustainability at Harvard. 2011. *Power Cogeneration Comes to Doubletree by Hilton.* Available: http://green.harvard.edu/power-cogeneration-comes-doubletree-hilton. Accessed 12/13/2013.

SWEEP. 2013. *DOE Awards $2.2 Million Grant to Southwest Energy Efficiency Project.* Available: http://www.swenergy.org/news/press/documents/PressRelease_DOEAward_CHP.pdf. Accessed 12/13/2013.

TAS. Undated. *Dell Children's Medical Combined Heat and Power Solution.* Available: http://files.harc.edu/Sites/GulfCoastCHP/CaseStudies/DellChildrenHospital.pdf. Accessed 12/13/2013.

TXCHPI. 2011. *Policy and Regulation.* Available: http://www.texaschpi.org/content/policy/policy.asp. Accessed 8/29/2013.

UGI Performance Solutions. 2012. *Reading Housing Authority: Glenside Homes.* Available: http://ugiperformance.com/documents/reading_housing_authority_case_study.pdf. Accessed 8/29/2013.

United States Energy Association (USEA). 2011. *Accelerating Combined Heat and Power Deployment.* Available: http://www1.eere.energy.gov/manufacturing/distributedenergy/pdfs/usea_chp_report.pdf. Accessed 12/13/2013.

U.S. DOE. Undated. *FEMP Factsheet: Landfill Gas to Energy for Federal Facilities.* Available: http://www.epa.gov/lmop/res/pdf/bio-alt.pdf. Accessed: 7/3/2008.

U.S. EPA. 2008a. *EPA CHP Partnership: County of Kauai Energy Extension.* Available: http://www.epa.gov/chp/partnership/partners/countyofkauaienergyextens.html. Accessed 8/29/2013.

U.S. EPA. 2008b. *Catalog of CHP Technologies. December 2008.* Available: http://epa.gov/chp/technologies.html. Accessed 8/29/2013.

U.S. EPA. 2011a. *Combined Heat and Power: Frequently Asked Questions.* Available: http://www.epa.gov/chp/documents/faq.pdf. Accessed 8/29/2013.

U.S. EPA. 2011b. *Opportunities for and Benefits of Combined Heat and Power at Wastewater Treatment Facilities. Market Analysis and Lessons from the Field.* Available: http://www.epa.gov/chp/documents/wwtf_opportunities.pdf. Accessed 8/22/2013.

U.S. EPA. 2013a. *CHP: Basic Information.* Available: http://www.epa.gov/chp/basic/index.html. Accessed 8/29/2013.

U.S. EPA. 2013b. *CHP: Efficiency Benefits.* Available: http://www.epa.gov/chp/basic/efficiency.html. Accessed 8/29/2013.

U.S. EPA. 2013c. *Methods for Calculating Efficiency.* Available: http://www.epa.gov/chp/basic/methods.html. Accessed 8/29/2013.

U.S. EPA. 2013d. *CHP: Reliability Benefits.* Available: http://www.epa.gov/chp/basic/reliability.html. Accessed 8/29/2013.

U.S. EPA. 2013e. *CT Long-term Loans for Customer-side DG.* Available: http://www.epa.gov/chp/policies/incentives/colowinterestloanprogram.html. Accessed 8/29/2013.

U.S. EPA. 2013f. *CHP: Environmental Benefits.* Available: http://www.epa.gov/chp/basic/environmental.html. Accessed 8/29/2013.

U.S. EPA. 2013g. *CHP: Economic Benefits.* Available: http://www.epa.gov/chp/basic/economics.html. Accessed 8/29/2013.

U.S. EPA. 2013h. *Lancaster County LFG Energy Project.* Available: http://www.epa.gov/lmop/projects-candidates/profiles/lancastercountylfgenergyp.html. Accessed 8/29/2013.

U.S. EPA. 2013i. *Biomass CHP Catalog: Chapter 3.* Available: http://www.epa.gov/chp/documents/biomass_chp_catalog_part3.pdf. Accessed 8/29/2013.

U.S. EPA. 2013j. *Procurement Guide: Selecting a Contractor/Project Developer.* Available: http://www.epa.gov/chp/documents/pguide.pdf. Accessed 8/29/2013.

U.S. EPA. 2013k. *EPA CHP Partnership: CHP Project Development: Operations and Maintenance.* Available: http://www.epa.gov/chp/project-development/stage5.html. Accessed 8/29/2013.

U.S. EPA. 2013l. *EPA CHP Partnership: CHP Project Development Process: Level 2 Feasibility.* Available: http://www.epa.gov/chp/project-development/stage3.html. Accessed 8/29/2013.

U.S. EPA. 2013m. *CHP Partnership dCHPP (CHP Policies and Incentives Database).* Available: http://epa.gov/chp/policies/database.html. Accessed 8/29/2013.

U.S. HUD. 2005. *CHP Guide #1.* Available: http://www.hud.gov/offices/cpd/library/energy/pdf/chpguide1.pdf. Accessed 8/24/2007.

Virginia DEQ. 2004. *Combined Heat and Power (CHP) and Distributed Energy Resources (DER) Summary and Synthesis of Regulatory and Administrative Impediments.* Available: http://www.deq.virginia.gov/innovtech/pdf/AR104.pdf. Accessed 8/24/2007.

We Energies. Undated. *Demonstration Projects: Lake Tower Office Building.* Available: http://www.we-energies.com/business_new/altenergy/laketower_demoproj.htm. Accessed 8/27/2007.

Winnebago. 2007. *Project Profile: Winnebago County Sherriff's Office.* Available: http://www.midwestcleanenergy.org/profiles/ProjectProfiles/Winnebago-CountySheriffs.pdf. Accessed 3/17/2008.

www.ingramcontent.com/pod-product-compliance
Lightning Source LLC
Chambersburg PA
CBHW081411170526
45166CB00010B/3295